REMOTE SURVEILLANCE
BY ELECTROMAGNETIC WAVES
for Air - Water - Land

by

DAG T. GJESSING

Chief Scientist
Remote Sensing Technology Program
Royal Norwegian Council for Scientific
and Industrial Research
Kjeller, Norway

ANN ARBOR SCIENCE
PUBLISHERS INC
P.O. BOX 1425 • ANN ARBOR, MICH. 48106

230 Collingwood, P. O. Box 1425, Ann Arbor, Michigan 48106

Library of Congress Catalog Card No. 77-85089
ISBN 0-250-40203-3

Manufactured in the United States of America

PREFACE

The last 20 years have been characterized by rapid progress in the field of radio science. At first the research was stimulated predominantly by radio communication interests. The search for better and more reliable communication channels has necessitated the consideration of a wide spectrum of geophysical factors affecting the propagation of electromagnetic waves. Atmospheric waves and turbulence are responsible for long-range, over-the-horizon scatter communication in the troposphere, the earth topography and vegetation have a dominating influence on ground wave communication, and the irregularity of the sea surface structure influences the propagation of radio waves that rely on reflections between the ionosphere and the earth's surface.

It was soon realized that radio waves constituted a diagnostic tool of high potential in the study of these terrestrial geophysical phenomena. A new field of radio science emerged, that of radio physics, radio geophysics and radio meteorology. Radio techniques were developed with the objective of studying the motion pattern of atmospheric winds, of temperature and humidity in the troposphere. Through these radio science methods a much improved understanding of the dynamic processes in our atmosphere was achieved. Similarly, inaccessible regions in space were studied with electromagnetic waves. Many years before modern space probes became available, the moon, the planets and the sun were studied extensively and successfully using radio waves.

Looking back on the last two decades of research in the field of radio science, the potential of electromagnetic waves as a diagnostic tool in the study of a very wide spectrum of phenomena is evident. In view of the historical background of radio science, it is not surprising that when the demand was accentuated for detailed information about the state of our environment as regards resources and pollution, powerful environmental surveillance methods rooted in the field of radio science emerged.

In developing this new field—applying radio science principles to environmental surveillance problems—the need arose for a unified treatment of these principles and their applications. This book is an endeavor to contribute to satisfying this demand. It represents an attempt to provide a coherent and brief outline of the underlying theory and of the basic principles behind present-day remote probing methodology. The book is intended for advanced students or scientists with good all-around knowledge in the field of radio science, but it does not assume previous acquaintance with this speciality. An effort has been made to produce a book that can be read from cover to cover and that reveals, if read in this way, some of the main ideas of radio science as applied to remote probing of environmental factors, pollution and resources.

In order to achieve this, severe limitations in scope had to be accepted. Some of the more ordinary topics of current interest are missing altogether and others will be found in skeletonized form only. To minimize this latter inconvenience, a bibliography of comprehensive articles is included.

Under the sponsorship of AGARD, a course on environmental surveillance by remote probing was given by the author in Ankara, Athens, Rome and Amsterdam during the fall of 1976. The experience gained on these occasions contributed constructively to the preparation of this book. It is a pleasure to acknowledge the initiative of the AGARD director, Mr. Olav Blichner. Likewise, I had the opportunity of preparing, together with Dr. Jack Nordo of the Norwegian Meterological Office, a state-of-the-art report on remote sensing for the Scientific Affairs Division of NATO. The help and encouragement of Dr. T. Kester of this agency is gratefully acknowledged.

Dag T. Gjessing

ACKNOWLEDGMENTS

Looking back upon the work in preparing this book, I feel a deep sense of gratitude to a number of people: colleagues, assisting staff and employers. To no one, however, do I owe a more sincere thanks than to my associates Terje Lund, Anton Kjelaas and Hugo Parr. Only through their expert assistance, their conscientious cooperation and their loyal support was it possible to complete this book during the time available. Questions helped to remove many obscurities, discussions gave stimulus and added momentum, and the companionship contributed to satisfaction and animation.

Among colleagues at the Norwegian Defence Research Establishment, my affiliation for more than 20 years, I am particularly indebted to Karl Holberg, Chief Scientist at the Electronics Division. His wisdom and his responsive and creative mind have been a continual source of inspiration and of crucial encouragement.

Likewise, I wish especially to thank Finn Lied, Director of the Norwegian Defence Research Establishment, under whose leadership I worked for many years. His sound judgment and his professional and scholarly example contributed substantially to my education and progress. I also wish to express my appreciation to Alan Waterman, Helge Ekre, Jack Nordo and Finn Bryn for many stimulating and clarifying discussions. Britt Tronvold, Kirstin Lillevold, Liv Borthen and Ruth Sand tolerantly and very ably handled the arduous task of typing the manuscript, and Marit Stoversten did all the art work for the figures; I wish to express my sincere thanks.

Finally, I wish to thank my family, and especially my wife Toril. The effort involved in preparing this book has taken a considerable amount of time away from them.

ABOUT THE AUTHOR

Dag T. Gjessing is Chief Scientist, Remote Sensing Technology Program, Royal Norwegian Council for Scientific and Industrial Research, Kjeller, Norway. He received his undergraduate degrees from Bergen University and London University, and his graduate degree in Geophysics from Oslo University. Dr. Gjessing has done research for Stanford University and the Norwegian Defence Research Establishment, Kjeller, Norway. He is a member of various international scientific organizations, among them the Norwegian Academy of Technical Sciences and the Norway Geophysical Society. He has contributed to over a score of international scientific journals and books on such topics as radiogeophysics, meteorology, measuring techniques (remote probing) and communication systems.

CONTENTS

CHAPTER 1

INTRODUCTION

The demand for powerful systems in the study of our environment is becoming increasingly urgent. The study of the earth's resources (minerals, vegetation, water) calls for observation methods that will provide information about the relevant parameters with adequate resolution in time and space. The environmental pollution problem leads to a specific demand: an assessment of the pollution burden to which our environment is subjected. Also, environmental surveillance methods are sought to establish and verify models on the basis of the assessments and predictions that can be made. Finally, observation methods are needed for the study of the dynamic processes that control the state variables, including the motion patterns of the sea and of the atmosphere.

To meet the demands for efficient surveillance systems we need methods based on fundamental principles, and we need the technology—the components—to implement the methods. During the last few years, significant technological advances have occurred in many important fields. We now have tunable electromagnetic sources that enable us to illuminate the object under investigation with electromagnetic waves, ranging from microwaves through infrared and optical waves into the ultraviolet region. Microwave receiver elements coupled to microcomputers are being developed; from these it is possible to determine the wavefront (amplitude and phase) over a large spatial region. The scattering of the illuminating wave from an object can thereby be investigated so that the object's geometrical shape can be determined. Electrooptical receiving systems, specifically superheterodyne receiver techniques, will soon give a dramatic improvement in receiver sensitivity (4-6 orders of magnitude) and frequency resolution, as was the case with microwaves some 30 years ago.

Low-cost and high-speed data processors (microcomputers) make it possible to perform the intricate and previously time-consuming computations necessary for detection/identification operations in real time. In order to apply existing techniques or techniques which are expected to become available in the near future, it will be necessary to develop the surveillance methods—architect the systems—to meet the specific demands.To achieve this, we need to gain an understanding of the basic principles involved and of the general merits of electromagnetic waves in the study of environmental factors.

It is the endeavor of this book to provide a basis for such an understanding. The aim is to present a unified and general treatment of the subject of remote sensing, basing the work on general and well-established results from radio science and on our knowledge of the interaction of electromagnetic waves with gases, liquids and solids. The book is organized as follows:

1. Fundamental Principles

These are included in the first section of the book, the purpose being to introduce some important radio science concepts and results.

- Fundamental principles in a remote study of the chemical composition of the atmosphere and of the land and sea surface, through knowledge about molecular structures.
- Basic principles related to the interaction of electromagnetic waves with surfaces of solids and liquids. Remote probing of the geometrical shape of objects on the earth's surface (vegetation, topography, wave structure and motion pattern of the sea and of the atmosphere).
- Basic propagation limitations in remote sensing. The influence of an atmospheric irregularity structure originating from turbulence on the resolution capability of a remote probing system.

2. General Theory of Remote Probing

- Methodology related to the measurement of molecular composition of reflecting or emitting surfaces and absorbing gases. Methods by which the chemical composition of the earth's surface can be determined by emission or reflection spectroscopy. This leads to specific methods and instrumentation, discussed in the last chapter.

- Concepts related to the determination of the geometrical structure of objects on the earth's surface by remote surveillance methods. Study of methods for the investigation of topography, of vegetation and of the wave structure and motion pattern of sea and atmosphere.

3. State-of-the-Art and Future Potential

Specific remote probing challenges on land, sea and air.

CHAPTER 2

FUNDAMENTAL PRINCIPLES IN THE STUDY OF THE MOLECULAR MICROSTRUCTURE

This section will focus on the molecule. Through a study of the molecular structure, information will be gained about the chemical composition of solids, liquids or gases in our environment. Our intention is to lay the foundation for microstructure investigations from remote platforms.

The tool at our disposal is electromagnetic waves. In studying liquid or solid surfaces, we illuminate the surface with electromagnetic waves and detect and analyze the reflected (scattered) signal. From this, it is possible to extract information about the molecular structure. Similarly, in the study of gases, we observe the electromagnetic waves that are subjected to absorption by the gas.

Fundamentally there are three physical principles that provide the basis for spectroscopic investigations of the molecular structure. *Absorption spectroscopy* is based on absorption of electromagnetic energy by the molecules at specific electromagnetic wavelengths. In the case of a gas, this absorption phenomenon is covered by the Beer-Lambert law, which states

$$P = P_i e^{-c\alpha_\lambda L}$$

Here P = intensity of electromagnetic wave after an absorption path of length L

P_i = incident intensity

c = gas concentration

α_λ = absorption coefficient for the wavelength λ

In the case of solid or liquid surfaces, the phenomenological problem is the same. When electromagnetic waves strike a surface, part of the energy is absorbed. The remaining energy is reflected back and is thus observable. Measuring the absorbed energy as a function of electromagnetic frequency, we obtain information about the molecular structure, as we shall see shortly.

Figure 2.1 illustrates the basic principles of absorption, reflection and emission spectroscopy. Note that the reflection and absorption modes of operation imply an active source of illumination, whereas in the case of emission the generator of electromagnetic waves is the body itself.

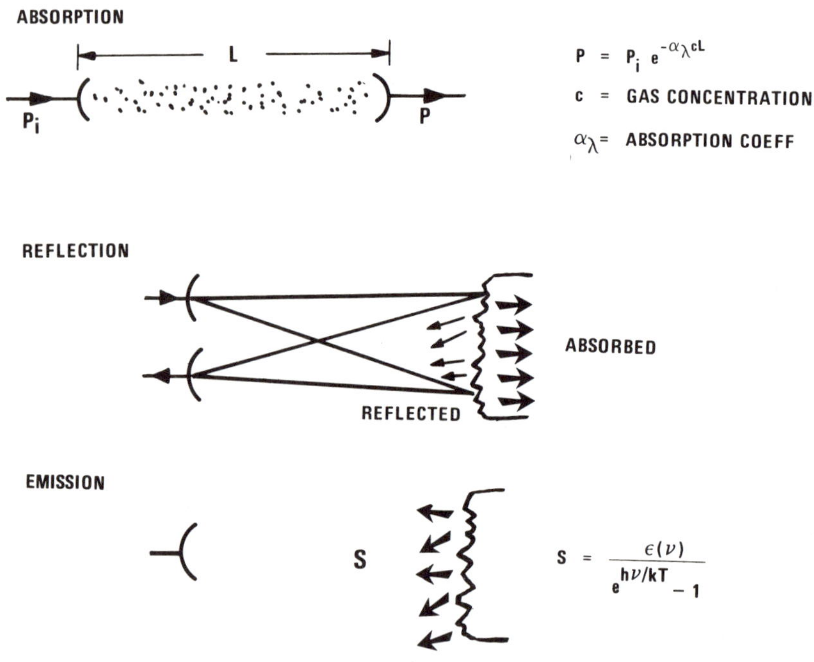

Figure 2.1. Study of molecular structure by absorption, reflection or emission spectroscopy.

Fluorescence spectroscopy is another technique with considerable potential. When molecules absorb electromagnetic energy as described above, part of the absorbed energy is reemitted at different frequencies. The frequency spectrum of the emitted fluorescent energy relative to that which stimulated the emission gives information about the molecular structure. Additional information for identification is obtained by measuring the rate at which the reemitted energy decays with time.

Raman scattering can, under certain conditions, provide significant detailed information about the molecular microstructure. If certain molecules (gases) are stimulated by a laser pulse of large intensity and short wavelength (ultraviolet radiation), each type of molecule will scatter light energy in all directions; the frequency of this reemitted light is frequency-shifted relative to the stimulating wave. Measurement of this frequency-shifted spectrum provides simultaneous information about the density of a large number of gases.

The basic absorption mechanisms of molecular spectroscopy will now be examined [1].

2.1 PURE ROTATIONAL SPECTRA

This category constitutes one of the principal classes of microwave molecular spectra. The spectra arise from the transitions between quantized rotational energies of molecules in single electronic ground states. If the molecule is excited by an electromagnetic field of a frequency ν corresponding to the energy difference between the ground state and the next level ($h\nu = E_1 - E_2$), the molecule will absorb this energy difference from the electromagnetic field, resulting in an absorption line at this frequency.

Consider a molecule as a rigid body rotating round an axis through the center of gravity. If the angular momentum around the axis of interest is I and the frequency of rotation is ω, the corresponding rotational energy will be given by

$$E_R = \frac{1}{2} I\omega^2$$

The quantum theory, however, requires the angular impulse $I\omega$ to be quantized in units of $h/2\pi$, where h is Planck's constant. This means that we can only have energies that are integral multiples of $h\nu$, leading to the following expression for the rotational energy E_R:

$$E_R = \frac{h^2}{8\pi^2 I} J(J+1) \tag{2.1}$$

where J is the rotational quantum number.

The selection rule for allowed transitions is that $\Delta J = \pm 1$. Therefore the absorption lines are equally spaced, with a spacing $h^2/(4\pi^2 I)$. The structure of rotational energy levels is shown in Figure 2.2, with the corresponding absorption spectrum. The main group

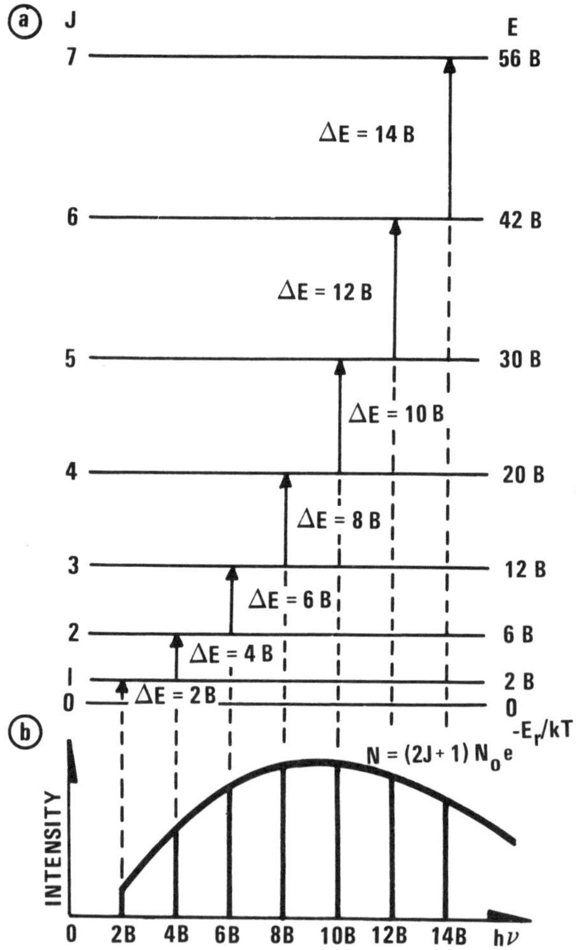

Figure 2.2 Intensity of line is determined by product of degeneracy and occupation probability. (a) Structure of rotational levels, with $B = h^2/(8\pi^2 I)$. (b) Resulting absorption spectrum for a molecular species.

of molecules giving rise to absorption lines within the range of microwave frequencies are the hydrogen halides. The only atmospheric gases offering absorption through the mechanism under consideration are SO_2, O_3 and H_2O.

2.2 INVERSION SPECTRA

Inversion may be described as the reflection of all the nuclei of a nonplanar molecule at its center of mass. In this way a new equilibrium configuration of the molecule is obtained, which in nonplanar molecules cannot be obtained from the original through any succession of simple rotations. To obtain the new equilibrium configuration, the molecule must pass through a potential barrier, which leads to a splitting of the vibrational energy levels. For NH_3 (ammonia) the corresponding inversion frequency in the ground vibrational state falls at about 24 GHz in the radio frequency region.

The action of the ammonia laser is based on this phenomenon. Other than ammonia, there is no important atmospheric gas possessing inversion spectra.

2.3 PARAMAGNETIC SPECTRA

These spectra are of great importance in connection with atmospheric absorption. The group of molecules that contain one or more unbalanced electron spins in the electronic ground state, such as O_2, NO, NO_2 and ClO_2, will be examined.

If the molecule is subjected to a magnetic field, the unbalanced spins will be weakly coupled to the magnetic momentum resulting from this field. Such a magnetic field will be generated, under certain circumstances, from the orbital motion of the electrons or from the end-over-end rotation of the whole molecule. The weak interaction between the spin magnetic moment S and the magnetic moment generated by the molecular rotation results in a splitting of each rotational level into (2S + 1) components, commonly known as the spin multiplets. Transitions between these triplets, brought about by an external magnetic field of frequency ν, will result in closely spaced absorption lines in (for O_2) the 4 to 6 mm region, and a single component in the 2.5-mm region.

2.4 VIBRATIONAL SPECTRA

Consider a molecule within which potential forces are acting; the atoms are bound together by elastic forces. If these atoms, having a finite mass, are excited by an alternating electromagnetic field, resonance will occur when the stimulating frequency equals that corresponding to the difference between two quantized vibrational energy levels.

Consider again two rigid atoms with masses m_1 and m_2, respectively. These two are tied together by forces that can be represented by a ball-and-spring system, where the restoring force obeys Hooke's law. The frequency of this simple, harmonic motion, oscillating ball-and-spring system is given by

$$\nu_o = \frac{1}{2\pi}\left(\frac{K}{\mu}\right)^{1/2} \tag{2.2}$$

where K = spring constant (Hooke's law)
μ = the reduced mass:

$$\mu = \frac{m_1 m_2}{m_1 + m_2}$$

Analogous to the case of rotational energies, we have that the energy at frequency ν is given by

$$E_V = (v + \frac{1}{2})\frac{h}{2\pi}\left(\frac{K}{\mu}\right)^{1/2} = (v + \frac{1}{2})\,h\nu_o \tag{2.3}$$

where v = 0, 1, 2, 3 . . . is the vibrational quantum number, and ν_o can be thought of as the "ground harmonic" of the oscillating mode.

The above example was given for two atoms. A complex molecule will have a large number of oscillating modes leading to numerous energy levels. However, Equation 2.3 still applies, with appropriate values of ν_o. It is possible to find the discrete energy levels, and thus the frequencies corresponding to maximum absorption, from knowledge about the molecular composition. From Equation 2.3, any particular mode will give a set of equally spaced energy levels. The vibrational spectra of importance all lie in the infrared range of frequencies.

2.5 ELECTRONIC SPECTRA OF POLYATOMIC MOLECULES

Molecules can absorb, or emit, radiation not only as a result of changes in their rotational and vibrational energies, but also as a result of changes in their electronic arrangement and, therefore, their electronic energy. The energy changes involved in a transition from one electronic state of a molecule to another are usually relatively large and correspond to radiation in the visible or ultraviolet regions.

The energy levels can be illustrated by one electron orbiting a nucleus. This quantum mechanics problem is solved using the Schrödinger equation. According to this, the electron can have energy levels given by

$$E_q = - \frac{N^2 e^2}{2a_o n^2} \qquad (2.4)$$

where n = 1, 2, 3 . . . the quantum number
Ne = charge of the nucleus
a_o = "Bohr radius" = $h^2/4\pi^2 m e^2$
$-e$ = charge of the electron
m = reduced mass of the electron

Quantitative results will help to illustrate the phenomenological discussion. Figure 2.3 [2] shows a theoretical absorption spectrum

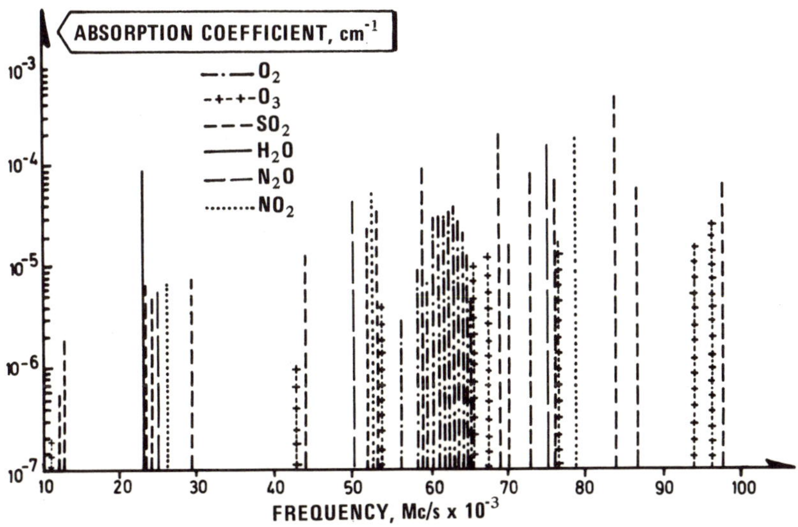

Figure 2.3 Microwave absorption coefficients of atmospheric gases in the frequency range 10 - 1000 GHz. After Ghosh and Malaviya [2].

for the most typical atmospheric gases at radio frequencies. The absorption coefficient at the peak resonance frequency is given by

$$\alpha_o = \frac{8\pi^2 \nu_o{}^2 N f}{3ckT} (\mu_{ij})^2 \cdot \frac{1}{\Delta\nu} \tag{2.5}$$

where N = number density of molecules
f = fraction of the total molecules present in the lower state
μ_{ij} = matrix element of the dipole moment μ, which gives rise to transition from i to j states
$\Delta\nu$ = half-width of absorption line
ν_o = peak resonance frequency
c = velocity of light

In terms of the conventional measure of attenuation dB/km, we have

$$\nu = 10^6 \alpha_\nu \log_{10}{}^e \quad \text{dB/km}$$

Let us consider the individual factors of Equation 2.5. N, the number density of molecules is given by

$$N = p/(kT)$$

where p = partial pressure of the absorbing gas
T = absolute temperature
k = Boltzmann's constant

Herzberg has shown that f, the fraction of molecules present in the lower states, is given by

$$f \sim \frac{1}{T} \quad \text{for diatomic and linear polyatomic molecules } (O_2 \text{ and } N_2O)$$

$$f \sim \frac{1}{T\,3/2} \quad \text{for symmetric and asymmetric top molecules} (H_2O \text{ and } O_3)$$

The linewidth $\Delta\ \nu$ varies with temperature and pressure as follows

$$\Delta\nu = (\Delta\nu)_{T=300K} \cdot P \cdot \left(\frac{300}{T}\right)^x \tag{2.6}$$

The value for x has been found experimentally (by Tinkham and Stradberg) for oxygen to be 3/4. For O_3, H_2O, N_2O, x has been assumed to be unity.

Let us then consider the shape of the absorption curve near resonance frequency. In terms of the resonance frequency ν_o, the half-width of the absorption line $\Delta\nu$ and the absorption coefficient at resonance, it can be shown that

$$\alpha_\nu = \alpha_0 \frac{\Delta\nu^2}{(\nu_0 - \nu)^2 + \Delta\nu^2} \cdot \frac{\nu^2}{\nu_0^{\,2}} \tag{2.7}$$

This expression is commonly known as the Lorentz factor, see Figure 2.4 [2].

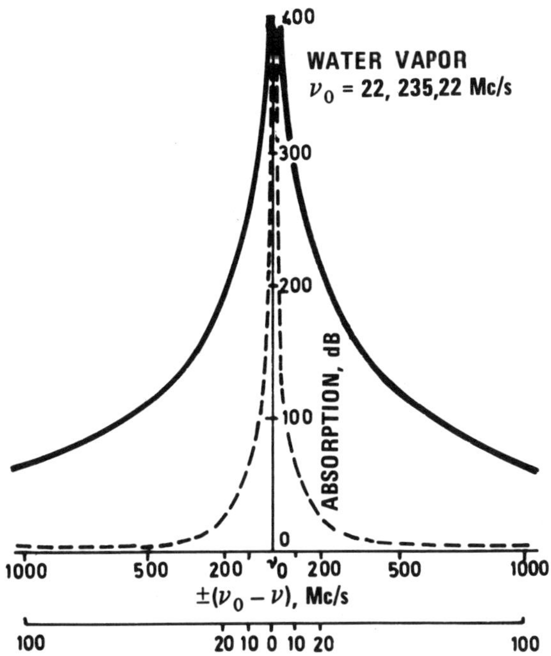

Figure 2.4 Total absorption around peak resonance freqeuncy for water vapor. After Ghosh and Malaviya [2].

Figure 2.5 shows the typical form of molecular absorption spectra arising from the different effects discussed.

To conclude the chapter on fundamental principles in the study of molecular structures, we note that from knowledge about the basic molecular structures, the absorption spectrum of a gas or the reflection spectrum of a liquid or solid surface can be computed, at least in principle. If, however, we are dealing with a mixture of many different molecules, the resulting spectra will be very

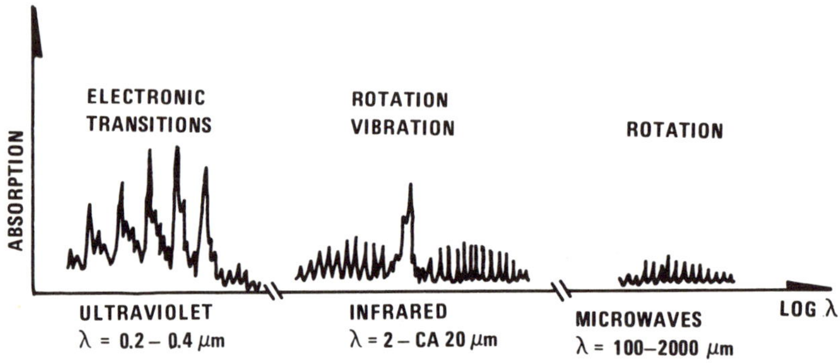

Figure 2.5 Typical form of molecular absorption spectra. Note that the vibrational transitions are split up by the rotational transitions.

complex. In practice, the composite spectrum can often be synthesized directly from knowledge about the absorption spectra (or reflection spectra as the case may be) of the individual constituents. The composite spectrum is obtained when the position of the individual absorption lines, the linewidth and the intensities for each constituent are known.

Using this approach, contributions from all the lines are added, taking into account factors such as line broadening and shifts in a gas due to thermal motion of and collision between molecules and local field effects in solids and liquids. Such comprehensive computations have been performed for the atmosphere by R. A. McClatchey and associates at the AFCRL laboratories in Massachusetts [see Reference 3].

Finally, we note that the remote sensing signature, the absorption spectrum or the reflection spectrum can be obtained from basic information about the constituents involved. As an example of an infrared absorption spectrum, the spectrum of CH_3I is given in Figure 2.6 for both the vapor and the liquid states. Note that the fine-scale structure is missing in the spectrum corresponding to the liquid state.

Figure 2.7 shows the theoretical monochormatic transmission over a 10-km horizontal path at sea level in the 10-μm region, assuming a midlatitude winter model atmosphere [3].

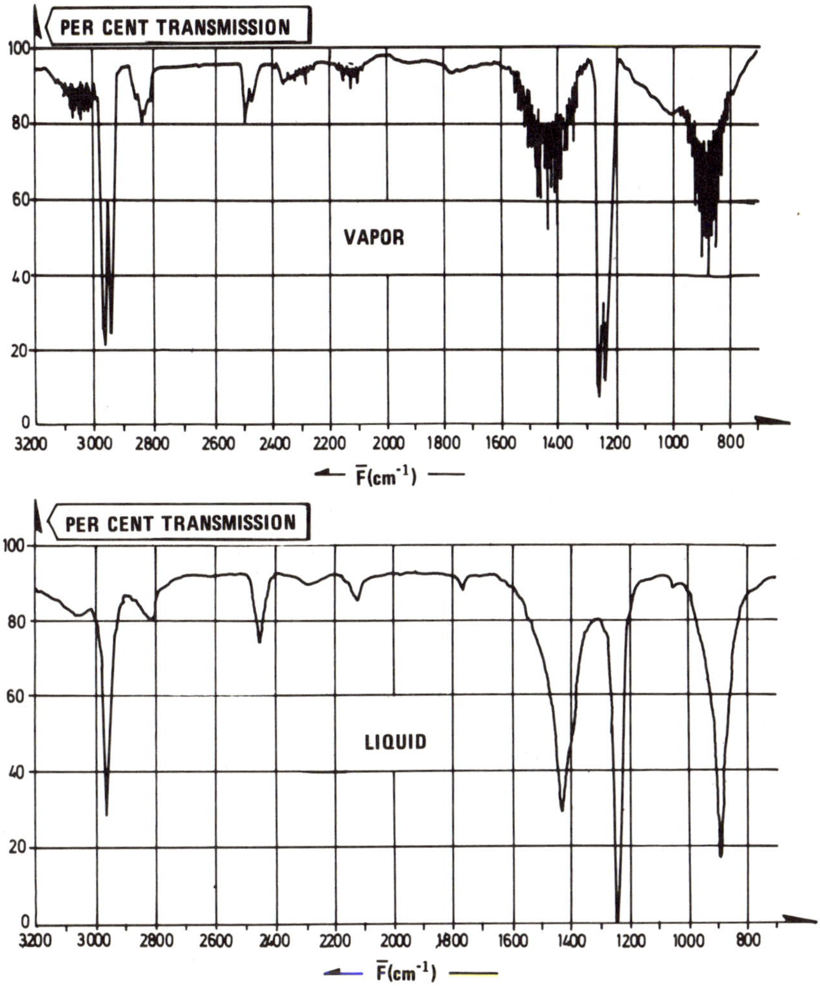

Figure 2.6 Example of an infrared absorption spectrum. The spectrum is for the liquid and vapor states of CH_3I. Note that the fine-scale structure is missing for the liquid state.

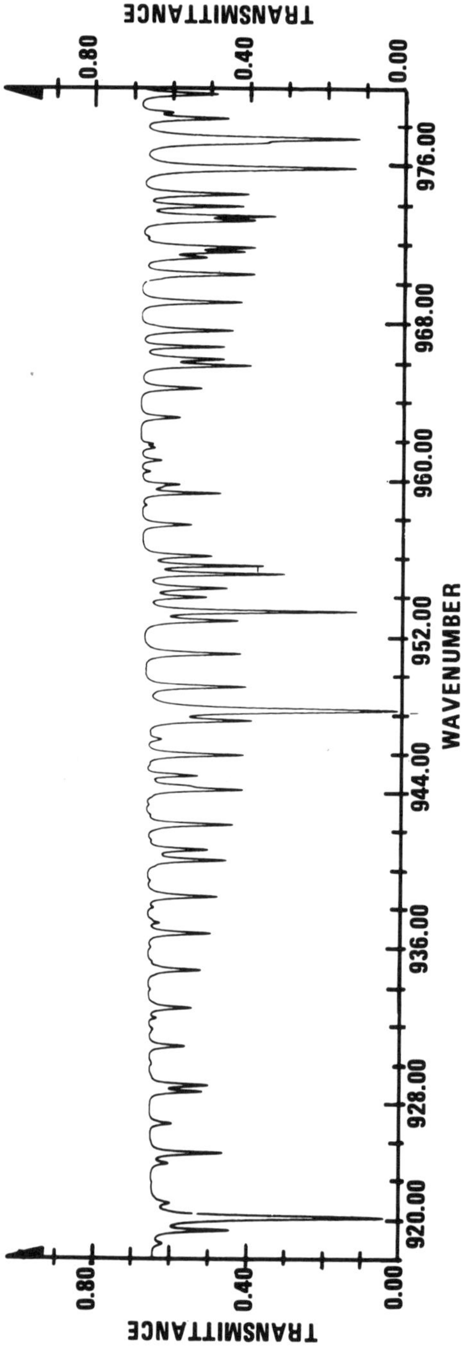
TRANSMITTANCE
0.80
0.40
0.00
920.00
928.00
936.00
944.00
952.00
960.00
968.00
976.00
WAVENUMBER
0.80
0.40
0.00
TRANSMITTANCE

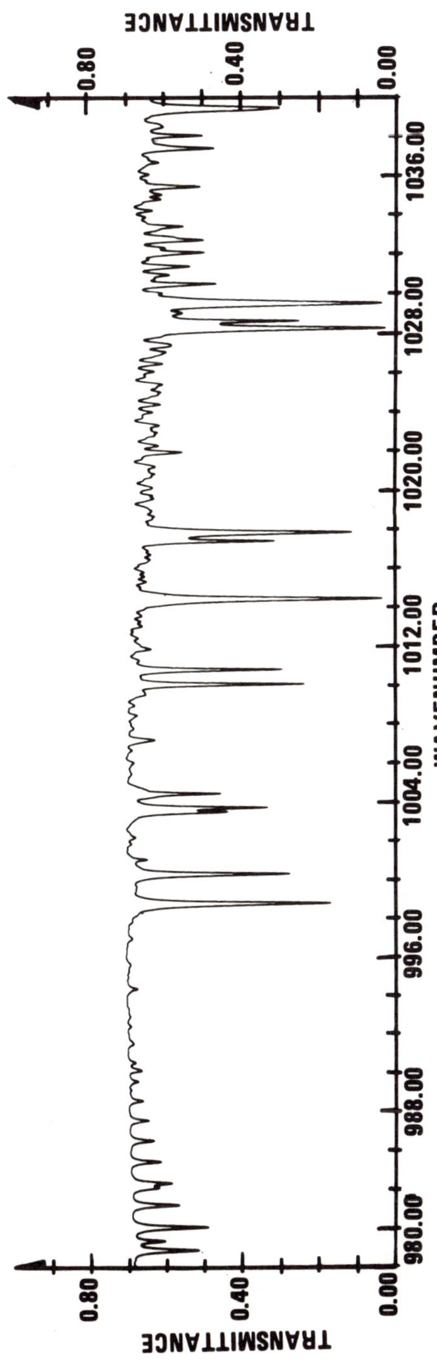

Figure 2.7 Atmospheric transmittance over 10-km horizontal path at sea level (10-μm region). After McClatchey *et al.* [3].

CHAPTER 3

REMOTE PROBING OF THE SURFACE MACROSTRUCTURE: ROUGHNESS OF THE EARTH'S SURFACE

Thus far we have been concerned with the microstructure molecular scale of absorbing, reflecting and emitting surfaces. We shall now concern ourselves with the macrostructure, the surface geometry, as revealed through surface irregularities, from the smallest, due to vegetation or fine-scale structured rock and sand, to those on a larger scale, due to topographic effects.

The question that arises is: Having at our disposal an illuminator (transmitter of electromagnetic waves) and a set of receivers capable of determining the properties of the wave reflected back from the object under investigation (ground surface, rock structure, vegetation), how should we proceed in order to obtain maximum information about the object's macrostructure? In practice, there are many ways of modulating a transmitter, and there are many specific antenna arrangements on the basis of which spatial resolution can be achieved. This chapter on basic principles will discuss the general properties of a wave scattered (reflected) from a rough surface. These are:

- correlation properties in space of the scattered wave (properties of the wavefront)
- correlation properties in the frequency domain of the scattered wave (bandwidth of the reflecting surface)
- temporal correlation properties of the scattered wave (power spectrum of the amplitude variation at a given radio frequency)
- amplitude distribution of the scattered wave (fading distribution)

Discussion of these characteristic properties of a scattered electromagnetic wave will form the basis for an understanding of the basic physical principles of remote probing that enable maximum information to be obtained about the reflecting surface from observations making use of electromagnetic waves (radio- to optical wavelengths).

3.1 SCATTERING/DIFFRACTION FROM IRREGULARITIES–BASIC RELATIONS

Our objective is to derive the basic expressions to be used in remote investigations of the earth surface macrostructure. However, before we attempt to synthesize the surface signature, we shall lay the basis for a detailed understanding of the physical principles involved by deriving the general equation governing diffraction and scattering processes. When dealing with the problem of scattering (reflection) from objects (scattering elements) that are distributed within a volume in space in a random or semirandom fashion, it is a matter of personal taste to base the discussion on a scattering process or to consider it as being a diffraction problem.

Radio propagation over natural obstructions has long been a subject of considerable interest. Classical knife-edge diffraction theory is used as an idealization of the physical problem. However, difficulties in obtaining an analytical solution, particularly when complex ground reflections are considered, have restricted most of the previous work to compilations of results from theory and measurements, *i.e.*, magnitude considerations only. These are of limited general interest in a presentation such as this. The knife-edge diffraction problem has been formally solved by several techniques. The early work by Sommerfeld, who based his analysis on the wave equation with appropriate boundary conditions, is perfectly rigorous and not restricted to small diffracting angles, but generalization to the multiple-reflection problem is not readily accomplished.

Fresnel and Kirchhoff based their work on an analytic formulation of Huygens' principle. The Fresnel-Kirchhoff method has the advantage of the simplicity of the ray-path concept; it is used almost exclusively in radio propagation literature. However, the method suffers from lack of rigor and from the appearance of the Fresnel integrals, which tend to obscure physical interpretations. In order to obtain practical solutions to the problem of diffraction in the presence of multiple reflections, a phasor summation of the Fresnel-Kirchhoff results is normally performed for the single-ray problem.

In a paper by Ratcliffe [4] a very ingenious and general method for solving complex diffraction problems is sketched. Ratcliffe shows that, in particular, if a one-dimensional diffracting screen is illuminated by a wave (wavelength λ), then the angular power spectrum $P(\theta)$ produced, expressed in terms of $\sin\theta$, is the Fourier transform of the spatial correlation $R_E(x)$ of the wavefront just as it leaves the screen, provided x is expressed in terms of λ.

Let us now derive the expression for the scattered/reflected/diffracted field from knowledge about the spatial distribution of the scattering elements and about the geometry. Note that we are considering the general case of three dimensions, not limiting ourselves to scattering from a surface only. Referring to Figure 3.1, consider a volume element $dv = dxdydz = d^3\vec{r}$, within the scattering volume V, this scattering volume being confined to the spatial region in the scattering strata illuminated by the transmitting antenna and "seen" by the receiving antenna. If the permittivity (refractive index squared) within the elementary volume differs by a small amount $\Delta\epsilon$ from the average value of the permittivity ϵ_o, the element of dielectric becomes polarized, giving rise to a dipole moment ($d\vec{P} = \Delta\epsilon\, dv\, \vec{E}_o$) when under the influence of an electric field $\vec{E}_o$. At a distance R from the scattering element, the dipole moment results in a polarization potential $d\vec{\pi}$ and, provided $k^2\vec{\pi} \gg \nabla\nabla\cdot\vec{\pi}$ (which requires $R \gg V^{1/3}$), the scattered field strength $\vec{E}_s \sim k^2\vec{\pi}$, where k is the wavenumber of the electric field. The scattered field resulting from the integral of elementary scattering elements is then given by

$$\vec{E}_s = \frac{k_s^2}{4\pi}\int_V \frac{1}{R}\vec{E}_o(\vec{r})\epsilon(\vec{r})e^{-j\vec{K}\cdot\vec{r}}\,d^3\vec{r}, \tag{3.1}$$

where the time factor $e^{j\omega t}$ has been omitted.

Here $\vec{K} = \vec{k}_o - \vec{k}_s$
$\vec{k}_o$ = wavenumber of incident field
$\vec{k}_s$ = wavenumber of scattered field

$|k_o| = |k_s| = 2\pi/\lambda$
λ = radio wavelength

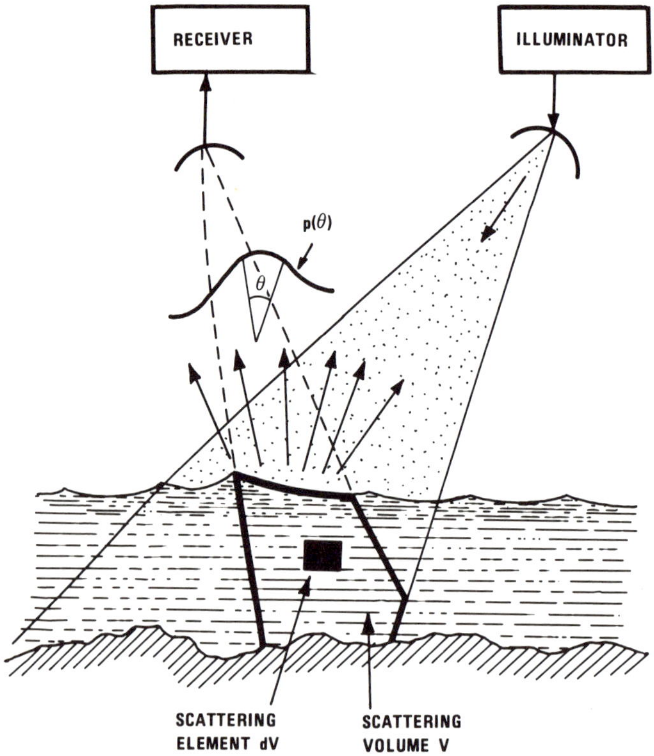

Figure 3.1 Scattering from irregularities distributed in depth on earth's surface.

Thus

$$|K| = \frac{4\pi}{\lambda} \sin \theta/2$$

where θ, the scattering angle, is the angle between $\vec{k}_s$ and $\vec{k}_o$. The vector $\vec{r}$ is the position vector of the scattering element, $\epsilon(\vec{r})$ is the space function of the normalized permittivity fluctuation $\Delta\epsilon/\epsilon_o$, and $\vec{E}_o(\vec{r})$ is the corresponding space function for the electric field.

Note that Equation 3.1, which is derived from Maxwell's equations, is general and does not consider the nature of the refractive index irregularities described by the function $\epsilon(\vec{r})$ or the irregularities in the electric field described by the $\vec{E}_o(\vec{r})$ function. The functions may be stochastic, in which case the refractive index field is conveniently described by the spatial autocorrelation function of the refractive index in the same way as the electric field $\vec{E}_o$ is described by its spatial autocorrelation function, or they may be ordered variations, *e.g.*, a horizontal layer through which the pertinent parameters vary in a systematic fashion as well-behaved functions.

From the basic Equation 3.1 for scattered field strength we see that there are two limiting cases:

1. If the field $\vec{E}_o(\vec{r})$ is constant within the scattering volume and the scattering volume is limited in extent (or varies slowly in space in comparison with the $\epsilon(\vec{r})$ function), then

$$\vec{E}_S = \frac{k^2 \vec{E}_O}{4\pi R} \int_V \epsilon(\vec{r}) e^{-j\vec{K}\cdot\vec{r}} \, d^3\vec{r} \qquad (3.2)$$

which states that *the scattered field $\vec{E}_s$ is proportional to the three-dimensional Fourier transform of the spatial variation in refractive index within the scattering volume V.*

2. If the $\epsilon(\vec{r})$ function is constant within the scattering volume and the field strength $\vec{E}_o$ varies, then the diffraction field $\vec{E}_D$ is the Fourier transform of the spatial variation in field strength within the scattering volume V.

$$\vec{E}_D = \frac{k^2 \epsilon}{4\pi R} \int_V \vec{E}(\vec{r}) \, e^{-j\vec{K}\cdot\vec{r}} \, d^3\vec{r} \qquad (3.3)$$

This is the same expression as that developed by Ratcliffe, referred to above [4]. See also References 5,6,7.

Returning to the classical approach; Huygens' principle is a well-known method for treating the propagation of waves. We shall now in a brief and semiintuitive way consider its mathematical background, showing its close connection with Green's theorem. The method is this: from each point of a given wavefront, at t = 0, we assume that spherical wavelets start. At time t, each wavelet will have a radius ct, and the envelope of these wavelets will form a new surface which, according to Huygens, is simply the resulting wavefront at this later time t. Thus, if the original wave was plane, it is easy to see that the later one will be a more distant plane, while if it is a sphere, the final wavefront will be a concentric sphere whose radius is greater by an amount ct. In either case this construction gives us the correct answer, agreeing with the more usual methods of computation.

We may look at our process in a slightly different way, not used by Huygens but developed later when the interference of light was being worked out. Suppose that, instead of taking the envelope of all the spherical wavelets, we consider that each of these wavelets has a certain amplitude, just as if our wave were being sent out by interfering sources of radio waves, with the resulting amplitude taken to be that of the actual wave. This process can lead essentially to the same result, and it is this that can be justified theoretically. As a further generalization, it is not necessary to take the original source to be a surface of constant phase; it can be any surface on which the amplitude and phase are specified.

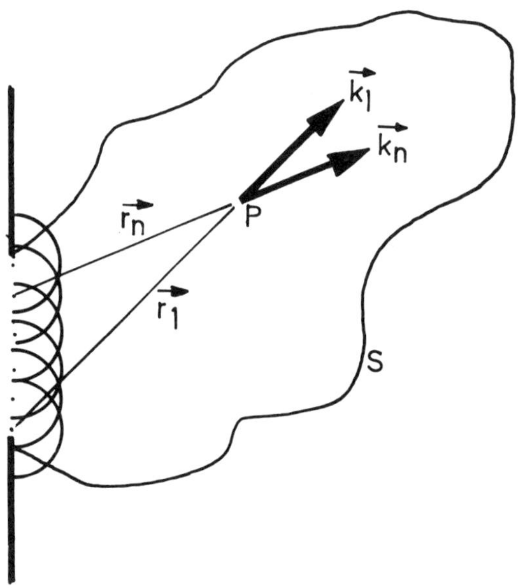

Figure 3.2 Diffraction geometry.

Our final result then is: The field at P, due to the waves transmitted from the slot, is obtained by taking an arbitrary surface around P through the slot and integrating over this surface. The contribution to the field at P of a small element of area dS on this surface equals the amplitude at P of a spherical wave starting from dS at such a time that it reaches P at a time t. That is, if the distance from dS to P is denoted by r, this wave is of the form

$$\frac{f\left(t - \frac{r}{c}\right)}{r}$$

Now, the contribution from a given wavelet must surely be proportional to the field at dS, which we may call $f(\vec{r},t)$, and to dS. Hence, for the field we have

$$f\,(\text{at P}) = \text{constant x} \iint \frac{f(t - r/c)}{r}\, dS$$

where the surface integral is taken over the surface surrounding P.

This suggests the solution of Laplace's equation by Green's method, where we have the value of a function ϕ at an interior point of a region where the surface integral over the boundary of $\nabla^2 \phi$ is zero. Performing a rigorous integration over the surface, setting the integral equal to zero except at the opening of the screen, we finally obtain an expression for the diffraction field of the form

$$F \sim \iint e^{-j\frac{2\pi}{\lambda}(r_1 + r_2)}\, dS$$

where r_1 is the distance from the transmitter (the illuminator) to the area dS, and r_2 the distance from dS to the point where F is measured.

So much for the classical diffraction theory. On the basis of the expressions above for scattered field strength $\vec{E}$, it then only remains to calculate the angular distribution of scattered power. To facilitate these calculations, let us rewrite Equation 3.3 above as follows [8]:

$$A(\vec{K}) = \int_V \vec{E}(\vec{r})e^{-j\vec{K}\cdot\vec{r}}\, d^3\vec{r} \tag{3.4}$$

so that the scattered/diffracted field strength is given by

$$\vec{E}_D = \frac{k^2 \epsilon}{4\pi R} A(\vec{K}) \tag{3.5}$$

The desired expression for the received power is obtained by multiplying $\vec{E}$ by its complex conjugate $\vec{E}^*$. Forming the product $A_i^*(\vec{K})\, A_j(\vec{K})$, we obtain the following:

$$A_i^*(\vec{K})A_j(\vec{K}) = \int\int \vec{E}(\vec{r})\,\vec{E}(\vec{r}+\vec{x})\,e^{j\vec{K}\cdot\vec{r}}\,e^{-j\vec{K}(\vec{r}+\vec{x})}d^3\vec{x}\,d^3\vec{r}$$
$$= \int d^3\vec{x}\,e^{-j\vec{K}\cdot\vec{x}} \int d^3\vec{r}\,\vec{E}(\vec{r})\,\vec{E}(\vec{r}+\vec{x})$$

The second integral is recognized as the spatial autocorrelation function for the electric field

$$R_E(\vec{x}) = \lim_{V\to\infty} \frac{1}{V} \int_{-V}^{V} \vec{E}(\vec{r})\,\vec{E}(\vec{r}+\vec{x})d^3\vec{r}$$

Hence

$$A_i^*(\vec{K})A_j(\vec{K}) = V\int R_E(\vec{x})\,e^{-j\vec{K}\cdot\vec{x}}\,d^3\vec{x}$$

But, according to the Wiener-Khintchine theorem [9], the power spectrum $\Phi\,(\vec{K})$ is given by

$$\Phi(\vec{K}) = \frac{1}{(2\pi)^3}\int R(\vec{x})\,e^{-j\vec{K}\cdot\vec{x}}\,d^3\vec{x}$$

and

$$P(\theta) \cong \vec{E}_D\,\vec{E}_D^* \cong \Phi(\vec{K}) \cong \int R_E(\vec{x})\,e^{-j\vec{K}\cdot\vec{x}}\,d^3\vec{x} \qquad (3.6)$$

Hence, the angular power spectrum $P(\theta)$ is the Fourier transform of the spatial autocorrelation of the electric field over the scattering volume. This important result will be used in the next section.

3.2 SPATIAL CORRELATION PROPERTIES OF ELECTROMAGNETIC WAVES REFLECTED FROM THE EARTH'S SURFACE; MEASUREMENT OF TRANSVERSE DISTRIBUTION OF SCATTERING SURFACE ELEMENTS BY REMOTE PROBING TECHNIQUES

Our attention is now focused on the transverse distribution of the scattering elements constituting the scattering object. In order to reveal this structure we shall use another characteristic of an electromagnetic wave, namely its spatial correlation properties. Classically, we know that there are three classes of antenna systems with a spatial resolution potential. These are:

- *Aperture antenna,* where the angular resolution $\beta = \lambda/D$, D being the aperture diameter and λ the radio wavelength. If R is the distance between the observation platform and the object to be analyzed, then the aperture antenna system has a resolution capability δx given by

$$\delta x = R\frac{\lambda}{D}$$

- *Array antenna,* which makes use of a set of antenna elements distributed along a baseline or in a plane. Each element is capable of measuring amplitude and phase of the wavefront at a point. On the basis of these point-observations of field strength, information about the target can be extracted. This will be the subject of the current section.
- *Synthetic aperture antenna,* which makes use of measurements of amplitude and phase over a baseline by shifting one sensor from one point to another, probing the wavefront sequentially in time (in series and not in parallel). In principle there is little difference between the linear array antenna and the synthetic aperture antenna [10]. See Figure 3.3.

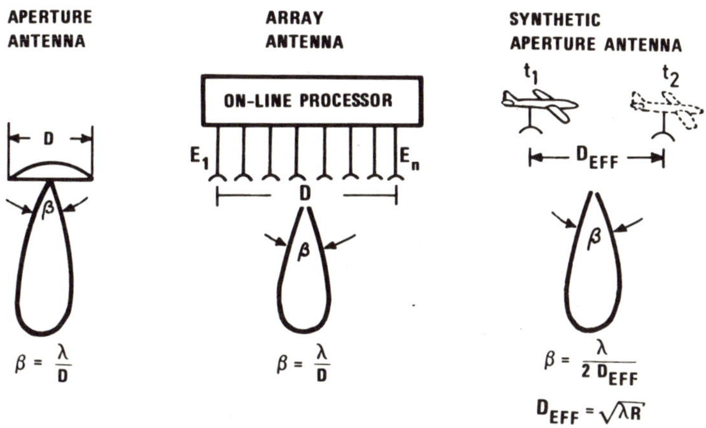

Figure 3.3 Basic principle related to three classes of high-resolution antenna systems.

If the object to be investigated is at rest and if atmospheric effects can be neglected, there is no difference in principle between the synthetic aperture antenna and the array antenna. However, as we shall see in the following chapter where we discuss the propagation limitations, irregularities in the atmospheric refractive index structure caused by turbulence impose particular limitations on the resolving power of a synthetic aperture antenna system. These aspects will be considered in Chapter 4.

We shall now analyze the phased array concept with a view to obtaining information about the transverse geometrical shape of the object by means of field strength measurements at n points along the baseline. Several approaches can be used. Since many of the derivations presented later in this book are essentially hinged on the relationship between the angle of arrival spectrum (angular power distribution) and the spatial correlation properties of field strength, it is probably justifiable to present two different approaches here.

The first approach is a direct consequence of the basic equations derived above. Referring specifically to Equation 3.6, which states that the angular power spectrum (the radiation pattern) emitted from a scattering/diffracting/reflecting area is the Fourier transform of the spatial autocorrelation function of field strength over this area, we see that this refers to the case where the area is considered as a radiator, *i.e.*, as a transmitting antenna.

In the case of a receiving antenna, the power reaching this array antenna is distributed as $P(\theta)$ over an angular region. Applying the inverse Fourier transform, it is self-evident that the $P(\theta)$ angle of arrival spectrum will give a spatial field strength autocorrelation over the array antenna, which is the Fourier transform of the $P(\theta)$ angular power spectrum. Thus, when our task is to measure the transverse distribution of the scatterers constituting the object of interest, we obtain information about this directly by using an array antenna measuring the autocorrelation function of field strength over the array. As is the case with any receiving antenna, the greater the required space resolution, the longer the antenna should be.

We shall return to this discussion shortly. Another approach, adopted by Hagfors [11], needs to be considered also. Suppose that the scattered field-wave impinging on the receiver is sensed by two small antennas displaced a distance ξ wavelengths along a direction perpendicular to the line joining the "transmitter," *i.e.*, the scatterers, and the receiver. At an instant t, the complex field strength from the direction θ (where θ is measured in a plane containing the two receiving antennas and the target from which the backscattered field originates) is taken to be $E(S,t)dS$ at the midpoint of the receiving antenna baseline. Here $S = \sin\theta$. The complex electromagnetic field induced in the two antenna elements from this direction is thus given by

$$de_1 \cong E(S,t)e^{-j\pi\xi S}\, dS \tag{3.7}$$

$$de_2 \cong E(S,t)e^{+j\pi\xi S}\, dS \tag{3.8}$$

provided the two antennas are identical so that the effective angular spectra can be assumed to be very nearly equal.

Integrating over all directions S, we obtain the total electromagnetic field in the two antennas

$$e_1 \cong \int E(S,t)e^{-j\pi\xi S}\, dS \tag{3.9}$$

$$e_2 \cong \int E(S,t)e^{+j\pi\xi S}\, dS \tag{3.10}$$

Note that the carrier frequency time variation is neglected and only the slow variation with time of the angular spectrum E(S,t) is retained.

The complex correlation of these two voltages is given by

$$\overline{e_1 \cdot e_2{}^*} \cong \int \vec{E}(S,t)\vec{E}^*(S,t)e^{-j\pi\xi(S+S')}dS\, dS' \tag{3.11}$$

$$= \iint \vec{E}(S,t)\vec{E}^*(S+\Delta S,t)e^{-j2\pi\xi S}\, e^{-j\pi\xi\Delta S}\, dS\, d(\Delta S) \tag{3.12}$$

where the bar refers to time averaging. If the angular correlation distance ΔS_{corr}, over which $\vec{E}(S,t)E^*(S+\Delta S,t)$ is much larger than zero, satisfies the condition

$$\Delta S_{corr} \ll \frac{1}{2\xi}$$

the integral can be simplified to

$$\overline{e_1 e_2{}^*} \cong \int \overline{|E(S,t)|^2}\, e^{-j2\pi\xi S}\, ds = \int |E(S)|^2\, e^{-j2\pi\xi S}\, dS \tag{3.13}$$

Physically, this means that we are interested in the angular power spectrum $|E(S)|^2$, the width S_o of which is determined by the size of ℓ_o of the scattering area that we want our array antenna to resolve. The size of our array antenna must therefore be of the order $\xi \cong 1/S_o$, so that the condition above becomes

$$S_{corr} \ll S_o \tag{3.14}$$

The normalized complex correlation of these two voltages is then

$$R(\xi) = \frac{e_1 \cdot e_2{}^*}{\overline{e_1{}^2}} = \frac{|E(S)|^2 \, e^{-j2\pi\xi S} \, dS}{|E(S)|^2 \, dS} \tag{3.15}$$

The above equation states that the spatial autocorrelation function $R(\xi)$ of field strength is the Fourier transform of the angular power spectrum (see References 4,5,6,7). Knowledge about $R(\xi)$ is thus a sufficient condition for determining the angular power spectrum $|E(S)|^2$. This is the same result as that obtained above [12].

Note that if ξ is measured along a horizontal direction, the angle of arrival spectrum is measured in the horizontal plane. If ξ is measured vertically, the angle of arrival spectrum of power ($|E(S)|^2 = P(\theta)$) is measured in the vertical plane. This angular distribution in the vertical plane $P(\theta)$ is proportional to the transverse vertical distribution $\sigma(x)$ of the scatterers constituting the target of interest. Without going into detail about this transverse distribution as measured through the angle of arrival spectrum and the effects of virtual scattering centers (giving effects such as glint), the first order relationship between the transverse distribution $\sigma(x)$ of the target scattering cross section and the angular power spectrum (angle of arrival spectrum) at distance R from the target is given by

$$P(\theta) \sim \sigma\left(\frac{x}{R}\right)$$

Thus, using as an example an object that gives constant scattering between certain limits Δx apart, the angular power spectrum $P(x/R)$ is found to be rectangular. The width is $\Delta x/R$, where Δx is the width of the object.

The Fourier transform of this rectangular angle of arrival spectrum is a $(\sin x)/x$ function, the width of which is given by

$$L \approx \frac{\lambda R}{\Delta x} \tag{3.16}$$

where λ is the wavelength of the radio wave.

This relationship is well known from antenna theory, from which we know that if L is the width of the illuminating field strength distribution (antenna aperture), the half power width of the resulting angular power spectrum is given by

$$\beta \approx \frac{\lambda}{L}$$

In the case of our "rectangular" target, since

$$\beta \approx \frac{\Delta x}{R} \tag{3.17}$$

the expression for L above follows directly.

Figure 3.4 illustrates the basic principles related to the determination of angular power spectrum on the basis of spatial field strength correlation measurements. By measuring the amplitude and phase of the wave scattered back from the object of interest at a number of points in a plane normal to the direction to the object, information about the transverse distribution of the scatterers constituting the object is obtained. Chapter 5, on general remote sensing methodology, presents some examples to illustrate the potential of the technique.

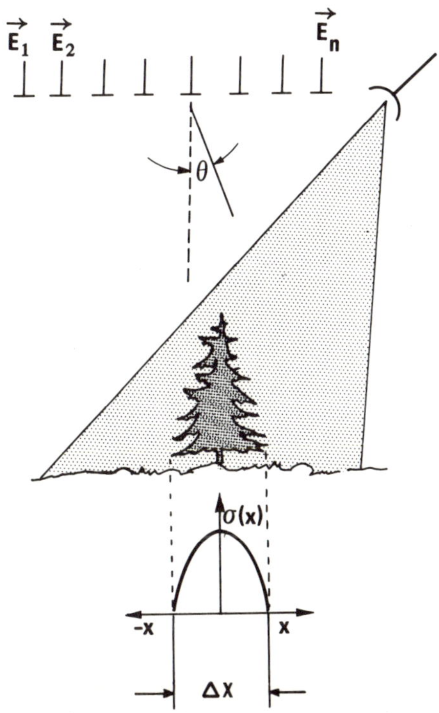

Figure 3.4 Basic principle for the measurement of transverse distribution of scatterers by spatial correlation measurement of field strength.

3.3 BANDWIDTH PROPERTIES OF ELECTROMAGNETIC WAVES REFLECTED FROM THE EARTH'S SURFACE; DISTRIBUTION IN DEPTH OF SCATTERING SURFACE [8]

In the first chapter we saw that in order to determine the molecular structure of the absorbing or reflecting material by measuring the absorption or reflection spectrum, the frequency of the transmitter (illuminator) had to be changed in some manner. There are two factors that influence the signal reflected back from the rough surface: (1) the molecular structure of the surface material, and (2) the geometrical shape (the roughness) of the surface.

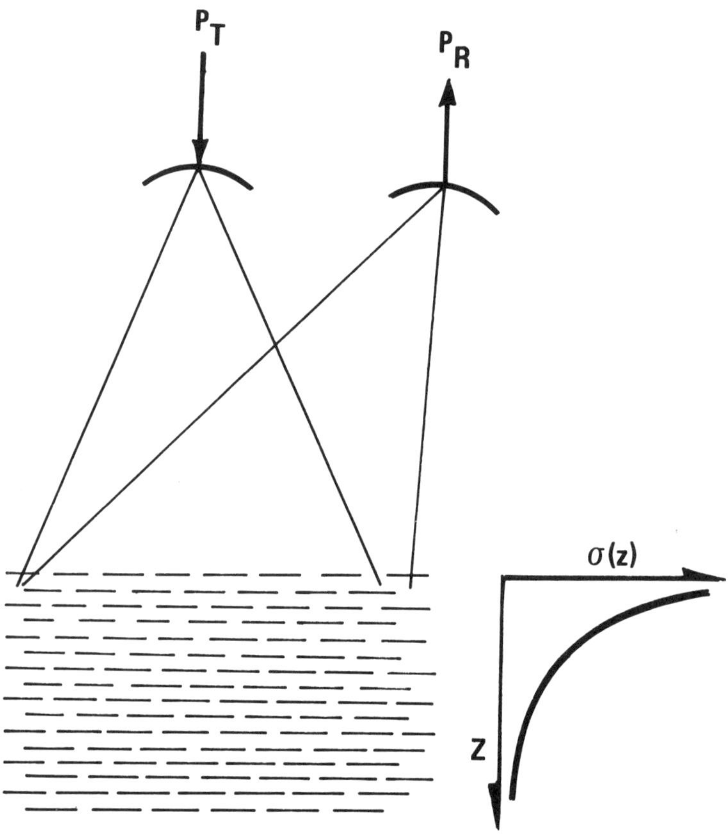

Figure 3.5 Geometry of the scattering in-depth process (correlation properties in the frequency domain).

In order to distinguish between the two effects, we must know both processes in some detail. There are three practical situations that warrant obtaining information about the correlation properties in the frequency domain of the reflecting surface:

- When the aim is to measure atmospheric gases by spectroscopic methods, it is practical to use the ground or the sea surface as the reflector, thus obtaining a monostatic measuring situation. Information about both the spectral and the geometrical effects of the reflector is then imperative.
- When information about the molecular surface structure is required, information about the influence of the surface roughness is imperative.
- Finally, as we shall see in the following, by measuring the correlation properties in the frequency domain resulting from a multifrequency illumination of the surface under investigation, information about the roughness of the surface is obtained (distribution in depth of the scattering surface).

Consider the case of diffuse (not specular) reflection (scattering) from a rough surface. The surface consists of a collection of scattering elements distributed in depth z, from which a set of waves is reflected back to the receiver. These waves will interfere (multipath fading) and the result is a signal with limited bandwidth, *i.e.*, with limited correlation properties in the frequency domain. We shall now calculate this correlation function (bandwidth of reflector).

A definition of "bandwidth" would help avoid the confusion that often arises when the term is used in relation to scattering processes. Consider the case in which several radio waves, each having different frequencies, are transmitted simultaneously. At the receiver, the power at each of these frequencies is measured as a function of time (*i.e.*, we measure $P_{F_1}(t)$, $P_{F_2}(t)$). If we take the instantaneous ratio of power at the various frequencies and integrate the ratio we form

$$\int \frac{P_{F_1}(t)}{P_{F_2}(t)}$$

We then obtain information about bandwidth. On the other hand, if we integrate the signal at either frequency over the appropriate time interval by forming

$$\frac{\int P_{F_1}(t)\,dt}{\int P_{F_2}(t)\,dt}$$

we do not obtain information about bandwidth but rather about "the wavelength dependence of the scatter circuit."

Forming the

$$\int \frac{P_{F_1}(t)}{P_{F_2}(t)}\,dt$$

is one way of obtaining information about bandwidth. Another way is to transmit a set of radio waves having different frequencies, as was done in the first case, letting all these frequencies be correlated in amplitude and phase. For example, this is achieved by amplitude modulating a carrier, thus obtaining two sidebands 2 F_{AM} apart, F_{AM} being the frequency of the modulating wave. These sidebands, obviously, are correlated in amplitude and phase. At the receiving end we pick up the two sidebands and correlate one with the other, *i.e.*, we form the cross-correlation function $R_{12}(\tau)$. The more narrow-banded the transmission channel, the poorer the correlation. (Transmitting many correlated waves spread over a frequency band enables us to find the complete autocorrelation function $R(\Delta F)$ in the frequency domain.)

If the scattering cross section σ associated with a scattering element distance z from the receiver is denoted $\sigma(z)$, the associated delay function of the received field is $P(\tau) \cong \sigma(z/c)$, where c is the velocity of light. Using analogous results from network theory, we obtain the following relationships: The Fourier transform (FT) of the delay function is known as the impulse response $F(\omega)$. Impulse response is thus

$$F(\omega) = FT(P(\tau))$$

Furthermore, the resulting power spectrum $W(\omega)$ is given by

$$W(\omega) = F(\omega)\,F^*(\omega) \tag{3.18}$$

Thus, if the distribution in depth of the scattering elements $\sigma(z)$ is known, the bandwidth of the reflecting surface $(\Delta\,\omega)$ given by $W(\Delta\omega)/W(o) = 1/2$ is obtained by a simple Fourier transformation of the $\sigma(z/c)$ function.

The scattering process from vegetation illustrates the essential points. We shall assume that the ground surface being illuminated has coniferous trees with needles distributed evenly in depth in such a way that the shadowing effect becomes progressively more dominant as the wave progresses. Let us, for the sake of simplicity, assume an exponential shadowing effect that distributes the illuminated scattering facets in an exponential manner in depth. This leads to a set of waves that will interfere arriving at the receiver. The result is a limited correlation bandwidth of the reflecting surface.

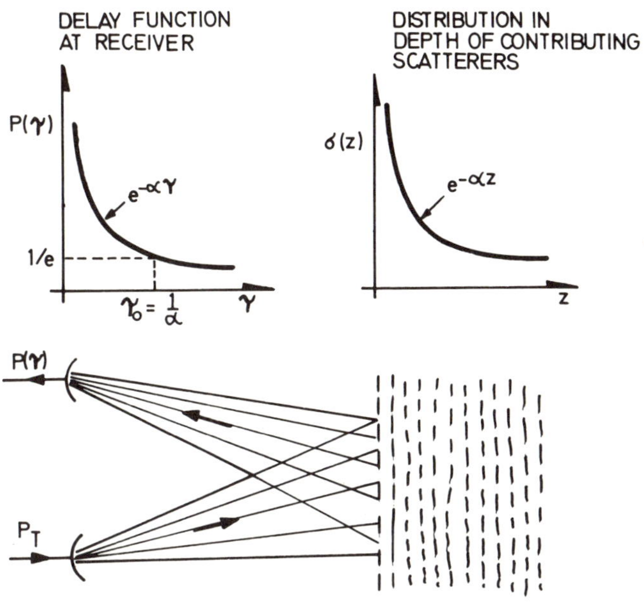

Figure 3.6 Geometry of the scattering process (correlation properties in the frequency domain).

This bandwidth can be calculated in the following manner. Since the distribution in depth of scattering cross section is assumed to be exponential, the delay function of these reflected waves is given by

$$P(\tau) = e^{-\alpha\tau}$$

The 1/e width of this delay function is given by

$$\tau_o = \frac{1}{\alpha}$$

The Fourier transform (FT) of the delay function is given by

Impulse response

$$\begin{aligned} F(\omega) &= \text{FT (delay function)} \\ &= \text{FT}\,(e^{-\alpha\tau}) \\ &= (\alpha + j\omega)^{-1} \end{aligned}$$

Thus the resulting power spectrum is [8]

$$W(\omega) = F(\omega)\,F^*(\omega) = (\alpha^2 + \omega^2)^{-1} \qquad (3.19)$$

The half power width $\Delta\omega$ of this spectrum is given by

$$\frac{W(\omega)}{W(o)} = \frac{1}{2} = \frac{\alpha^2}{\alpha^2 + \Delta\omega^2} \qquad (3.20)$$

in other words:

$$\Delta\omega = \alpha = 1/\tau_o$$

Hence, if the 1/e penetration depth of the electromagnetic wave is $c\tau_o = \Delta z$ meters, the bandwidth (width of power spectrum) is given by

$$\Delta F = \frac{c}{2\pi\Delta z}$$

The bandwidth discussion above may appeal to the radio engineer. However, to the scientist familiar with statistics and signal analysis it may be more meaningful to express the results in terms of the correlation properties of the signal scattered back from the rough terrain surface. Specifically, we shall now calculate the autocorrelation function in the frequency domain $R(\Delta\omega)$. The voltage V_1 of the signal scattered back at frequency ω is given by

$$V_1 = F(\omega) = (\alpha + j\omega)^{-1} \qquad (3.21)$$

Similarly, the voltage V_2 at frequency $(\omega + \Delta\omega)$ is given by

$$V_2 = F(\omega + \Delta\omega) = [\alpha + j(\omega + \Delta\omega)]^{-1} \qquad (3.22)$$

The normalized complex autocorrelation of these two voltages is then given by

$$R(\Delta\omega) = \frac{\int_{-\infty}^{\infty} (\alpha + j\omega)^{-1} \, [\alpha - j(\omega + \Delta\omega)]^{-1} \, d\omega}{\int_{-\infty}^{\infty} (\alpha^2 + \omega^2)^{-1} \, d\omega} \tag{3.23}$$

Solving this integral we derive the following expression for the modulus of the autocorrelation function:

$$R(\Delta\omega) = [1 + (\frac{\Delta\omega}{2\alpha})^2 \,]^{1/2} \tag{3.24}$$

Defining now the correlation distance in the frequency domain to be the halfwidth of the autocorrelation function, we have

$$\Delta\omega_{1/2} = 2\alpha\sqrt{3}$$

$$\Delta F = \frac{\sqrt{3}}{\pi} \frac{c}{\Delta z} \text{ Hz} \tag{3.25}$$

For easy visualization, this simple expression is plotted linearly in Figure 3.7. Note that the distribution in depth of the scatterers

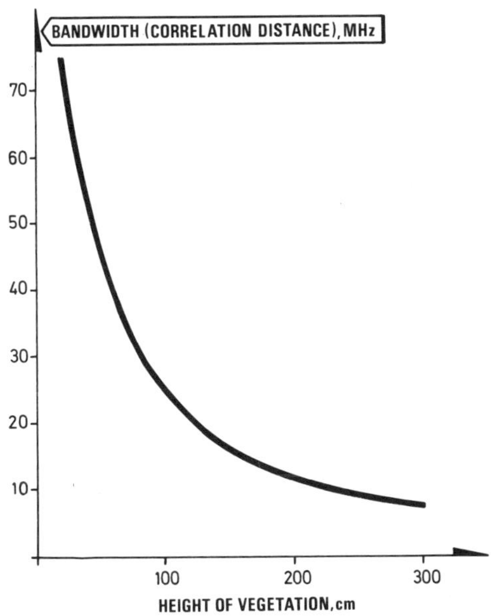

Figure 3.7 Bandwidth of a reflector consisting of vegetation with exponential distribution in depth of the scattering elements (exponential shadowing effect).

is assumed to be exponential, so that the height of the vegetation shown as the abscissa of Figure 3.7 is the 1/e width of the exponential depth distribution function. Had we assumed a different distribution in depth of the vegetation, we would have obtained a different curve relating bandwidth and height of vegetation. However, the inverse relationship would always exist, whereas the constant of proportionality would differ (see Chapter 5).

In conclusion, in order to study the fine-scale properties of vegetation, we illuminate the field of interest with a set of radiowaves (in the microwave region). The waves should be mutually correlated (*e.g.*, sidebands resulting from amplitude modulation) and have frequencies distributed over a few gigaherz. Correlating the electromagnetic field strength received at each frequency gives information about the detailed structure. If we assume a given structure form, *e.g.*, a given type of grass structure, and our interest is confined to the size of the grass, illumination at two frequencies only is required. If the object of interest is large trees, the distribution of our radio frequencies could be limited to the MHz range. In the chapter dealing with remote probing methods, some examples illustrating this feature will be given.

3.4 TEMPORAL CORRELATION PROPERTIES OF THE SCATTERED WAVES; WIDTH OF THE POWER SPECTRUM OF THE AMPLITUDE SCINTILLATIONS

We have seen that in order to extract the information of interest —the concentration of a particular chemical compound—from a background of noise, signal processing methods involving temporal integration are necessary. The question then arises: How do the signals resulting from backscattering from the terrain, as discussed above, decorrelate with time? This property of the signal determines the length of the integration time that we have at our disposal. In yet other applications, such as in radar study of sea state, the fluctuating return from the rough surface itself constitutes the "signal." Whether or not the terrain return is considered noise or signal, it is very important to understand its temporal correlation properties and its statistics, treated in Chapter 4.

There are several factors that determine the temporal correlation properties of the scattered signal. These have been studied extensively over many years [5,8,9,13-15]. We shall present the main results. In order to ensure an understanding of the physical principles involved, we shall derive the results from first principles. Consider a scattering element within the spatial region (the scattering volume) contributing to the scattered signal. The wave incident on this has a wavevector $\vec{k}_1$, whereas that of the scattered wave is $\vec{k}_S$. If this scattering element is in motion relative to the observation platform—the transmitter/receiver—then the wave scattered back by the scattering element with velocity $\vec{V}$ is subjected to a Doppler shift F. This frequency shift is given by the familiar relationship

$$F = \frac{1}{2\pi}(\vec{k}_1 - \vec{k}_S) \cdot \vec{V} \tag{3.26}$$

$$= \frac{1}{2\pi}\vec{K} \cdot \vec{V}$$

Assuming a monostatic radar situation as depicted in Figure 3.8, we measure pure backscatter such that the angle between $\vec{k}_1$ and $\vec{k}_S$ is 180° and the vector difference $\vec{K}$ is directed along the direction of propagation. The magnitude of the vector difference $\vec{K}$ is

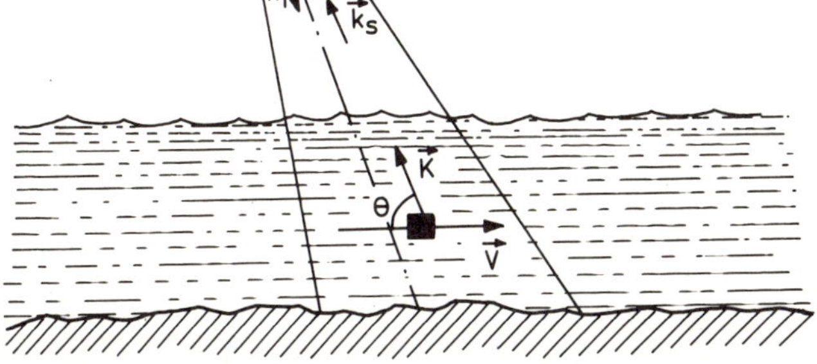

Figure 3.8 The scattered signal is Doppler shifted as a consequence of the motion of the scattering element relative to the radar.

then $4\pi/\lambda$. Thus, from one signal scattering element we have a Doppler shift F given by [16,17]

$$F = \frac{1}{2\pi} \cdot \frac{4\pi}{\lambda} V \cos\theta \tag{3.27}$$

$$= \frac{2V}{\lambda} \cos\theta$$

where θ is the angle between $\vec{K}$ and $\vec{V}$. However, we have many scattering elements contributing at any one given time. The result is that a set of waves reaches the receiver. These will have different Doppler shifts, phases and amplitudes. This situation leads to amplitude scintillations, or fading. There are two limiting cases that should be considered.

1. Influence on the scintillation spectrum of the velocity distribution of the scattering elements

Let us assume that the beamwidth β is so small that the difference vector $\vec{K}$ is constant within the scattering volume. The width ΔF of the Doppler spectrum causing scintillation is then

$$\Delta F_V = \frac{1}{2\pi} \vec{K} \cdot \delta\vec{V} \tag{3.28}$$

$$= \frac{2\delta V}{\lambda} \cos\theta$$

when the width of the velocity distribution is δV.

The correlation time τ is inversely proportional to ΔF. As an example, let us consider the scattered signal from a patch of the ocean surface, using a 3-cm (X-band) stationary radar. The radar senses the ripples (capillary waves) that have an oscillation frequency around 20 Hz. Decorrelation will be even faster, since each ripple oscillation only has to proceed a fraction of 2π in phase before the total backscattered signal from all ripples has changed. In fact, a scintillation frequency of about 100 Hz is observed. Thus, every 10 ms the radar sees essentially a "new" patch of ocean, uncorrelated to the previous one. In the presence of swell, the signal is also modulated by the much longer gravity waves, and a second correlation time of several seconds appears. (Remote probing of the ocean is treated in some detail in Chapter 5).

2. Influence on scintillation spectrum of the spatial distribution of the scatterers; constant speed of the scattering elements

Let us now consider the case where all the scattering elements are moving at the same velocity $\vec{V}$. If, in addition, the difference wavevector $\vec{K}$ is also constant, there will be no Doppler broadening, but merely a constant Doppler shift causing no scintillation. Using a finite beamwidth β on our radar antenna, however, the direction of the wavevector $\vec{K}$ will vary within the limits of β. The corresponding Doppler broadening is then

$$\Delta F_K = \frac{1}{2\pi} \Delta\vec{K} \cdot \vec{V}$$

$$= \frac{2V}{\lambda} (\cos\theta - \cos(\theta + \beta))$$

which by expansion and approximation yields

$$\Delta F_K = \frac{2V}{\lambda} \sin\beta \sin\theta \qquad (3.29)$$

In the above we have not assumed any structure in the distribution of scatterers over the illuminated area. If there is such a structure, say a periodic grating with period ℓ_o, or if we have an irregular sea surface with scale length ℓ_o (the width of the spatial autocorrelation function describing the irregularity structure is ℓ_o), then this may show up in the scintillation spectrum of our radar.

Consider now two limiting cases. First assume that the beamwidth β is very narrow so that the irregularity structure of scale length ℓ_o is resolved. Flying with constant speed V over the irregularity structure, a fluctuation component f given by

$$2\pi f = V \cdot K$$

i.e.,

$$f = \frac{V}{\ell_o} \qquad (3.30)$$

is obtained [15].

Similarly, consider the case where a large number of irregularities are illuminated by a plane wave. Let the typical scale length be ℓ_o. As Ratcliffe originally pointed out in connection with the ionosphere [4], the autocorrelation function of the complex amplitude (phase included) will then have a scale length ℓ_o at *all distances* from the scattering region. In the same way as above, the scintillation frequency will be $f = V/\ell_o$, provided both phase and amplitude are measured. The result that the spatial correlation length of the complex amplitude is independent of distance may seem surprising at first. Physically, it can be motivated by noting that the "scale length information" is carried by the signals from scatterers lying far from the normal viewing direction, so that the observer is far outside of the first Fresnel zone of most of the contributing scatterers.

In conclusion then, when calculating the combined effect of all these contributions to the scintillation spectrum, several factors have to be taken into consideration. The relative importance of these depends on the geometric and geophysical conditions prevailing. For example, consider an airborne radar. Let the resolution cell on the ground be determined in azimuth angle ϕ, and at a grazing angle θ (see Figure 3.9). The use of Equation 3.28 yields the following spreads in relative velocities:

$$\Delta V_{azimuth} = V \cos\theta \left[\frac{\beta}{2} \sin\phi + \frac{\beta^2}{2} \cos\phi \right]$$

$$\Delta V_{radially} = V \cos\phi \sin\theta \tan\theta \frac{c\tau}{2R} \qquad (3.31)$$

The corresponding scintillation frequency is approximately given by

$$\Delta f \approx \frac{\Delta V_{tot}}{\lambda}, \quad \Delta V_{tot} = [(\Delta V_{az})^2 + (\Delta V_{rad})^2]^{1/2} \qquad (3.32)$$

Numerically, consider an X-band radar flying at 1500-m altitude at a speed of 100 m/s, with a beamwidth of 5.3° and a grazing angle of 5.6°. The azimuth Doppler decorrelation then dominates, giving a scintillation frequency of 7 Hz straight forward, and 300 Hz in the side direction, corresponding to Doppler beats between different scatterers within the resolution cell. Depending upon the particular application, this Doppler decorrelation can be considered both as a nuisance and a blessing. On the one hand, the motion of the scatterers will mask slower time variations, making their study impossible. On the other hand, it will permit independent measurements spaced in time by $1/\Delta f$, thus making averaging easy

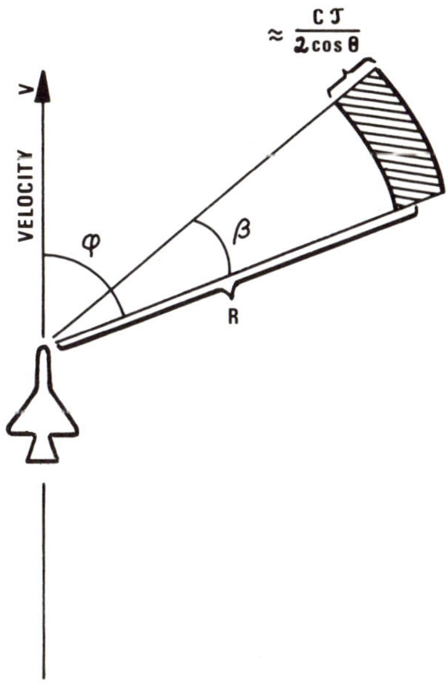

Figure 3.9 Notation for the Doppler width analysis. Beamwidth is β, pulse-length τ, azimuthal angle ϕ. Grazing angle relative to terrain is θ (not shown).

and effective. In that case, the degree to which the signal-to-noise ratio can be improved by an integration process is determined by the decorrelation time of the signal relative to that of the noise. In our example, pointing the radar sideways gives many more independent samples than in the forward direction, assuming the signal to be the same in the two directions.

3.5 AMPLITUDE DISTRIBUTION OF ELECTROMAGNETIC WAVES SCATTERED FROM A ROUGH SURFACE (VEGETATION)

We have seen that the signal received after diffuse scattering from a rough surface is a result of many elementary waves stemming from an integral of scattering elements. There are two different multipath mechanisms that give rise to fading:

1. If the scattering elements (leaves, branches) are in motion, systematically or randomly, this motion gives rise to a Doppler frequency shift of the individual elementary waves. The power

density at the receiver is thus a result of a number of electromagnetic waves that have random mutual phase, different frequencies and, in general, different amplitudes. This situation leads to fading.

2. The scattering elements are at rest, but the refractive index integral along the ray path from the scattering element to the receiver varies. Thus the relative phase angle of the integral of elementary waves varies. This too leads to fading. As regards the amplitude distribution of the received signal, several workers [*e.g.*, 13] have shown that this distribution is independent of the shape of the Doppler spectrum and of its position on the frequency axes. Indeed, the amplitude distribution is affected only by the number of incoherent scattering elements present. Thus, a knowledge about the amplitude distributions gives a rather meager contribution to the general understanding of the structure of scattering elements contributing to the receiver field. If the number of scattering elements is large (larger than 5-6), then the amplitude distribution will closely resemble a Rayleigh distribution [13].

CHAPTER 4

PROPAGATION LIMITATIONS IN REMOTE SENSING

The two preceding chapters presented methods on which the geometrical structure (macrostructure) and the chemical composition (the micro- or molecular structure) of a scattering surface can be determined by means of frequency-modulated electromagnetic waves. Specifically, in analyzing the solid object or the absorbing gas, three properties of the electromagnetic waves, as measurable at the observation platform, are utilized:

- The correlation properties in the frequency domain of the electromagnetic wave are studied. A set of waves with different but correlated frequencies is transmitted, and the correlation properties of the waves reflected back from the scattering surface are analyzed as to frequencies.
- The correlation properties of the scattered electromagnetic field in space are studied. The surface is illuminated with a plane wave, and the correlation properties of the scattered field are studied spatially, using an array of detectors capable of measuring amplitude and phase of the reflected wave.
- Finally, the temporal correlation properties are measured.

However, in travelling from the observation platform to the scene of interest (gas volume in space, ground surface to be analyzed), the waves must pass through an atmospheric path over which the properties determining the propagation performance vary. Specifically, these parameters are the refractive index of the atmosphere and its variability in space and time (wind velocity). In order to make sure that the influence of the propagation medium is kept under control and does not constitute a major hindrance to a detailed study of the environmental factors of interest, comprehensive information about the effects of the propagation medium is needed.

In this context it should be mentioned that radiometeorologists for more than two decades have used radio waves to study the detailed atmospheric structure. They have even used much the same general technique as those treated here, suggesting that the influence of atmospheric factors on the electromagnetic wave is an important one and must be taken into consideration when other environmental factors are studied. It also suggests, however, that a substantial amount of work has already been completed in this field, and, consequently, solutions to many of the problems already exist. We shall now, with reference to this work, give some of the more important results.

4.1 INFLUENCE OF ATMOSPHERIC EFFECTS ON THE SPATIAL CORRELATION PROPERTIES OF AN ELECTROMAGNETIC WAVE

Irregularities in the atmospheric refractive index structure lead to multipath phenomena and delay variations and a spread in the angle of arrival when an electromagnetic wave passes through the irregular transmission medium, causing a loss in bandwidth and reduced spatial correlation of waves. Some theoretical results, which are well-confirmed experimentally, and which give information about the amplitude covariance of the field at points separated in a plane normal to the direction of propagation, will be discussed. However, before presenting the accurate results of comprehensive calculations, we shall, as in the case of scattering mechanisms above, give some quantitative results for the purpose of ensuring an understanding of the basic physics involved.

The simple geometrical sketch of Figure 4.1 illustrates two extreme paths through which the electromagnetic waves can travel from the transmitter T to the receiver R. One is the shortest direct way from T to R, the other is via a path which is $\lambda/2$ wavelengths longer than the direct route. The result of these two waves is the vector sum of two signals with a 180° difference in phase causing destructive interferences. To the first order,

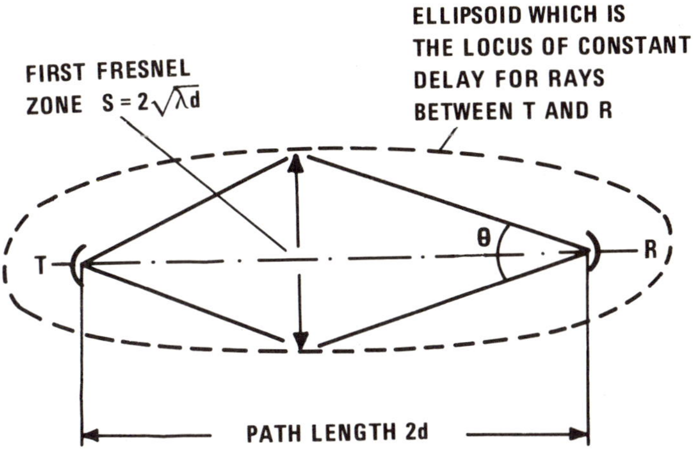

Figure 4.1 The geometry of line-of-sight propagation.

therefore, we would expect the width θ of the angle of arrival spectrum at the receiving point to be given by

$$\tan^{-1} \theta/2 = \frac{\sqrt{\lambda d}}{d}$$

$$\theta \cong 2\sqrt{\lambda/d} \qquad (4.1)$$

Knowing the angular power spectrum, *i.e.*, the beamwidth, the correlation distance in a plane through the point of the receiver normal to the line T-R can be calculated. We have shown in Chapter 3 that the spatial correlation function of the field is the Fourier transform of this angular power spectrum [12]. If this power spectrum is a $[(\sin x)/x]^2$ function, then the Fourier transform is a rectangular function. If the width of this correlation function of field is L, then we have the following relationship between the half-power width of the beam $\theta_{1/2}$ and the correlation distance L

$$\theta_{1/2} = \frac{0.88\lambda}{L} \qquad (4.2)$$

where λ is the wavelength of the electromagnetic wave.

In passing, note that this is the same expression as that relating antenna beamwidth $\theta_{1/2}$ to the antenna aperture diameter L. This is not surprising since the antenna radiation pattern [the $P(\theta)$ function] is the Fourier transform of the autocorrelation function of the field across the antenna aperture.

Therefore, from Equations 4.1 and 4.2 above, we have

$$\frac{0.88\lambda}{L} = \frac{2\sqrt{\lambda d}}{d}$$

i.e.,

$$\text{correlation distance } L = 0.44\sqrt{\lambda d}$$

Thus, the correlation distance of the field transverse to the line of propagation is comparable with the first Fresnel zone. This is shown in Figure 4.2. Note that in this figure the normalization factor is $\sqrt{\lambda d}$, where d is the path length, not half path length as in Figure 4.1. Hence we shall have to multiply our approximate correlation distance by $\sqrt{2}$ when referring to Figure 4.2.

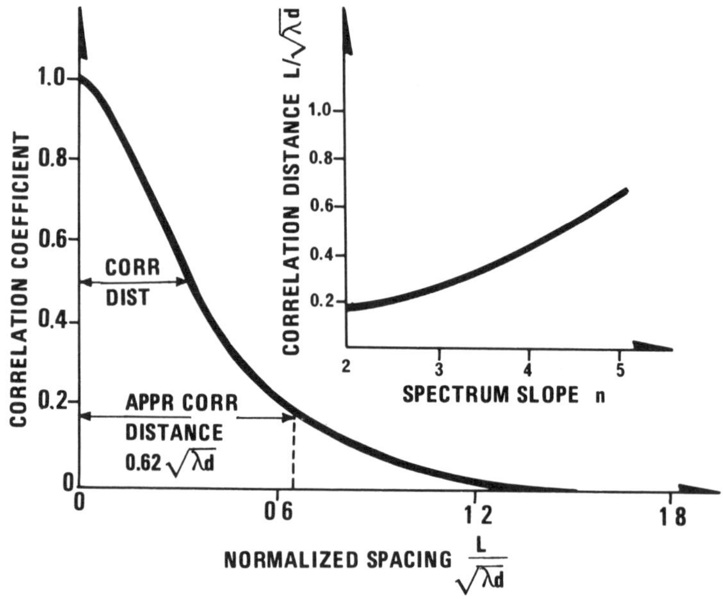

Figure 4.2 Spatial correlation of the field. Normalization factor $\sqrt{\lambda d}$ refers to path length d. From Lee and Harp [18], Kjelaas [19].

Let us now discuss this aspect in some greater degree of detail, referring to more rigorous calculations. The index of refraction, which through spatial variations gives rise to the scattering and refraction phenomena, is normally very small and can be written in the form

$$n(\vec{r},t) = 1 + n_1(\vec{r},t)$$

where the temporal dependence is mainly due to the wind. Under such conditions there is a direct relationship between the spatial variation of the refractive index and the temporal variations in a point. This relationship is known as the Taylor frozen flow hypothesis, stating that $2\pi f = VK$, where f is the frequency of the temporal variations at a point, V is wind speed and K is spatial wavenumber.

In the following, therefore, we are interested in the variations only; we are interested in the second moment, the covariance $< n_1(\vec{r}) n_1(\vec{r}\,')>$, where the outer brackets denote ensemble averaging. If we furthermore assume that the random refractivity fluctuations locally are statistically stationary, the covariance of n_1 can be written

$$< n_1(\vec{r})\, n_1(\vec{r}\,') > = \int\int_{-\infty}^{\infty}\int d^3\vec{K}\, \Phi(\vec{R},\vec{K})\, e^{j\vec{K}\cdot\vec{\rho}} \tag{4.3}$$

where $\Phi(K)$ is the wavenumber spectrum, as introduced in Chapter 3, of the spatial refractive index fluctuations.

The following relations are also introduced

$$\vec{R} = \frac{\vec{r} + \vec{r}\,'}{2}$$

$$\vec{\rho} = \vec{r} - \vec{r}\,'$$

The multidimensional integral in Equation 4.3 above can be reduced to single integrals. For the two-dimensional problem at hand with line-of-sight propagation, we have

$$\int\int e^{j\vec{K}\cdot\vec{\rho}}\, \Phi(\vec{R},\vec{K})\, d\vec{K} = 2\pi \int K\, \Phi(R,K) J_o(K,\rho)\, dK \tag{4.4}$$

This expression is recognized as that corresponding to the diffraction of a plane wavefront by a circularly symmetric aperture.

Referring to Lee and Harp [18] it can be shown that the covariance between waves with different wavenumbers k_1 and k_2 (where $k = 2\pi/\lambda$) is given by

$$C_f(\rho) = 4\pi^2 k_1 k_2 \int_o^{L_p} ds \int_o^{\sqrt{k_1 k_2}} dK \cdot K \cdot \Phi(K)\, J_o\left(\frac{sK\rho}{L_p}\right) \cdot \sin\left(\frac{K^2 s(L_p - s)}{2\, k_1\, L_p}\right) \cdot \sin\left(\frac{K^2 s(L_p - s)}{2\, k_2\, L_p}\right) \tag{4.5}$$

where ρ = spatial separation of receivers between which the correlation is measured

L_p = length of transmission path

s = elementary distance along the path L

K = wavenumber of refractive index irregularities ($K = 2\pi/\ell$ when ℓ is the spatial size of the irregularities)

$\Phi(K)$ = spectrum of refractive index irregularities. This is of the form $\Phi(K) \cong K^{-n}$, n being an atmospheric parameter, which varies with conditions (stability) in the atmosphere [20]. The median value of n appears to be close to 4.

J_o = Bessel function of zero order

The trigonometric terms in Equation 4.5 may be interpreted as a geometric weighting function. This general expression can be simplified if we use one frequency only, thus putting $k_1 = k_2$. The expression for the amplitude covariance between waves at different points in space with orthogonal separation ρ relative to the direction of propagation is then

$$C(\rho) = 4\pi^2 k^2 \int_o^{L_p} ds \int_o^{k} dK \cdot K \cdot \Phi(K) J_o\left(\frac{sK\rho}{L_p}\right) \sin^2\left(\frac{K^2 s(L_p - s)}{2kL_p}\right) \qquad (4.6)$$

This function is plotted in Figure 4.2 for a Kolmogorov spectrum when $\Phi(K) \cong K^{-11/3}$. The inserted figure shows how the correlation distance (the halfwidth of the covariance function) varies with spectrum slope n (where n is given by $\Phi(K) \cong K^{-n}$). Note that the approximate results agree well with the curve obtained for rigorous computations.

Let us now investigate what happens if we transmit two frequencies simultaneously and correlate the signals at the two frequencies in one point at the receiving end.

4.2 INFLUENCE OF ATMOSPHERIC EFFECTS ON THE FREQUENCY CORRELATION PROPERTIES OF AN ELECTROMAGNETIC WAVE

Having obtained the general expression, Equation 4.5, which is generalized to give the covariances in both space and frequency, the derived expression for the amplitude covariance versus frequency separation f_1/f_2 follows directly. The frequency covariance function for the Kolmogorov spectrum $\Phi(K) \cong K^{-11/3}$ is plotted in Figure 4.3 for $\rho = 0$. As an insertion in the figure giving

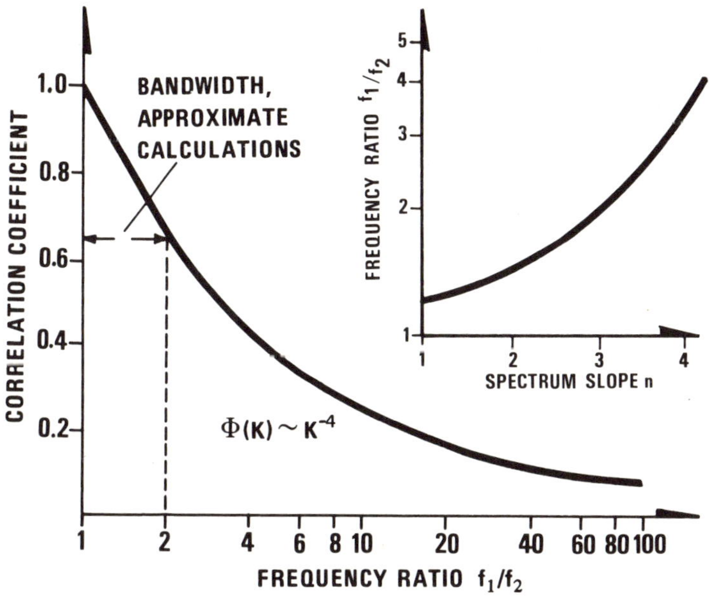

Figure 4.3 Amplitude covariance vs frequency separation f_1/f_2. After Lee and Harp [18]. The insertion shows the influence on bandwidth of the slope n of the refractive index spectrum ($\Phi(K) \cong K^{-n}$). Bandwidth is defined as λ_1/λ_2 for correlation coefficient = 1/2.

the frequency covariance function $R(\Delta F)$, a diagram illustrating the relationship between bandwidth and atmospheric structure is given [19]. Note that the structure parameters are the familiar n parameter ($\Phi(K) \cong K^{-n}$).

Finally, let us calculate the correlation distance in the frequency domain (the bandwidth) using a simple, approximate approach. From Chapter 3, we learned that the bandwidth function (autocorrelation function in the frequency domain) is obtained by Fourier transforming the delay function.

For simplicity, let us again assume that the delay function is of the (sin x)/x form with a half-power width $\Delta\tau = (\lambda/2)/c$. The bandwidth function, *i.e.*, the frequency transfer function $P(\Delta F)$ or the correlation function in the frequency domain $R(\Delta F)$, would then be a rectangular function, the width of which is

$$\Delta F = \frac{1}{\Delta\tau} = \frac{2c}{\lambda}$$

= half-frequency of the electromagnetic wave

These were the approximate results. We see that these too fit well on the curve giving the comprehensive results.

4.3 INFLUENCE OF ATMOSPHERIC EFFECTS ON THE TEMPORAL CORRELATION PROPERTIES OF AN ELECTROMAGNETIC WAVE

We shall now consider the temporal correlation properties of the transmission medium. Information about these is important if our measurements are not performed instantaneously in time. As an example, if we are to study the molecular structure of an atmospheric gas using a multifrequency illumination system, the temporal correlation properties of the atmosphere become important, if measurements are not performed at the same instant in time at all the pertinent frequencies.

We suffer a decrease in temporal correlation because of the following facts. Consider a "frozen" atmospheric refractive index irregularity structure. If the frozen transmission medium is kept at rest, all the multipath waves keep to the same trajectories, resulting in no changes with time. If, however, the frozen block is moved in a direction normal to the direction of the electromagnetic wave propagation, conditions will change. The faster the medium is moved, the more rapid become the changes and the shorter is the temporal correlation time. Without elaborating on this mathematically, it can be shown (which intuitively appears to be obvious) that information about temporal covariance properties is obtained directly from the spatial covariance function through the simple relationship

$$t = \frac{\ell}{V}$$

Here t = temporal correlation distance
ℓ = spatial correlation distance
V = wind velocity

Thus the temporal amplitude covariance function is obtained directly by scaling the spatial covariance function.

4.4 IMPLICATIONS OF AN IRREGULAR ATMOSPHERIC REFRACTIVITY STRUCTURE

The fact that there is a transmission medium between the observation platform and the specimen to be investigated imposes limitations on bandwidth and on the correlation properties of the electromagnetic wave in both space and time. This makes it necessary to take the atmospheric factors into consideration when designing the measuring system. For the purpose of illustrating this, we shall give two examples.

4.4.1 The Influence of Atmospheric Refractive Index Irregularities On the Resolution Performance of a Radar

In the following example, we shall focus our attention on the resolving power of a sidelooking radar under conditions of atmospheric limitations. There are several schemes by which a synthetic aperture antenna can be made to give optimum resolution. In the case where the phase curvature departs less than $\lambda/8$ from the plane over the observation baseline, the effective (synthetic) aperture is given by

$$D_{eff} = 2\sqrt{\lambda R} \tag{4.7}$$

so that the width of the synthetic "antenna diagram" is given by

$$\beta = \frac{\lambda}{2\sqrt{\lambda R}} \tag{4.8}$$

and the corresponding target resolution Δ at the distance R is given by

$$\Delta = R\beta = \frac{\sqrt{\lambda R}}{2} \tag{4.9}$$

If the field of the arriving wave is not correlated over the synthetic aperture $2\sqrt{\lambda R}$, a loss in resolving power is suffered.

In the preceding section, we developed expressions for the spatial covariance function of amplitude fluctuations in terms of the pertinent atmospheric parameters. Referring to Figure 4.2, we see that when the slope n of the refractivity spectrum $\Phi(K) \cong K^{-n}$ varies, so does the spatial correlation distance of the amplitude fluctuations. A typical distribution function for the refractive index structure parameter n, in Figure 4.4, shows that n varies substantially [8,21,22].

Using this probability distribution, we find the corresponding distribution for the correlation distance of the field from Figure 4.2. From antenna theory, we know that only if the field is correlated in amplitude and phase over the antenna aperture D will the resolving power Δ of the antenna of distance R from the object be given by

$$\Delta = \frac{\lambda}{D} R$$

If the correlation distance of the field is less than the antenna aperture D, then the resolving power will be reduced. The degree to which the resolution is reduced by atmospheric effects depends in practice on the signal processing capability of our receiving system.

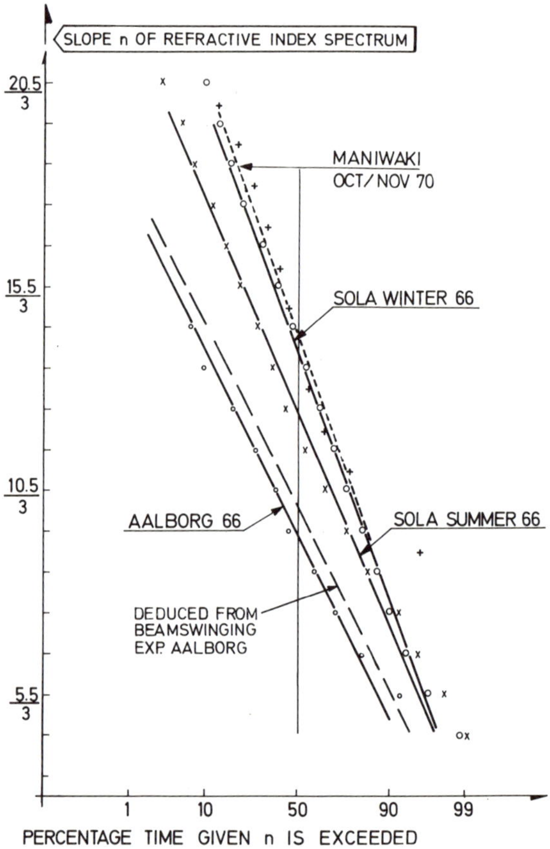

Figure 4.4 Distributions of the slope of the spectrum of refractive index irregularities as deduced from radiosonde observations and beam-swinging experiments.

In this example we shall define the resolution capability as

$$\Delta' = \frac{\lambda R}{\text{correlation distance}}$$

In the preceding section we gave the relationship between the correlation distance and the slope n of the refractive index spectrum. This correlation distance is defined as the antenna spacing required for the correlation coefficient of amplitude fluctuations at the two antennas to drop to 1/2.

In Figure 4.5 the probability distribution of the resolution capability Δ′ is plotted for a 10-km path and compared with that which would have been achieved in the case of no refractive index

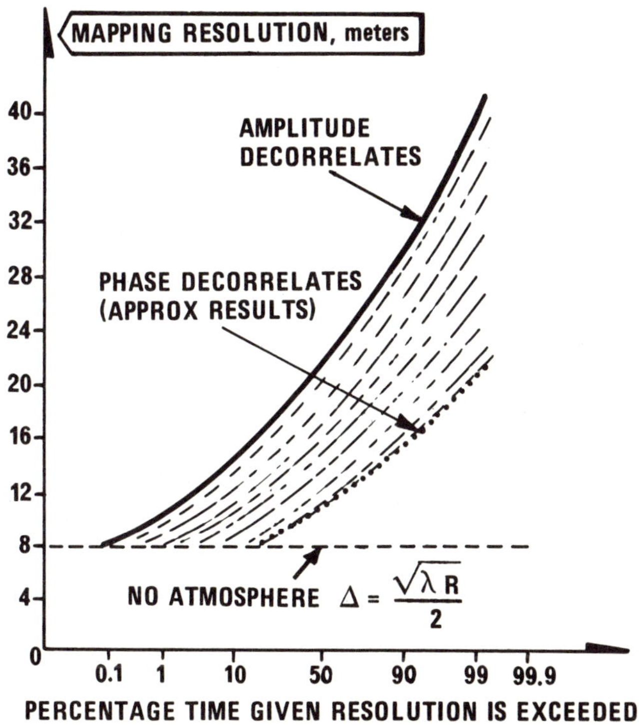

Figure 4.5 Probability distribution of target resolution calculated from radiosonde data and compared with the resolution that would have been obtained with a synthetic aperture antenna under conditions of no refractive index irregularities. Note that the curve for phase is approximate only.

irregularities in the transmission medium [10]. Note that these results are based on the correlation distance of amplitude only. To complete the picture, we have also to consider the fluctuations in phase and the spatial correlation properties of these. Obviously, if the phase variations caused by atmospheric effects are comparable with the wavelength and not correlated over the aperture of the synthetic antenna, we shall suffer a dramatic loss in resolution capability. Thus, in the probability distribution of spatial resolution capability given in Figure 4.5 we shall need a second curve, namely that determined by the correlation distance of the phase fluctuations. Between the resolution determined by a limited correlation distance of amplitude and that determined by phase, there will be a "gray zone."

When the baseline of our synthetic aperture antenna is first increased, the amplitude scintillations will play a role and introduce some blurring effects. Increasing the baseline still further makes the lack of phase coherence noticeable. This will greatly influence the resolution capability of our synthetic aperture radar system. However, the correlation distance of phase is determined not only by the slope n of the refractive index fluctuations, but also by the scale length L_o. The outer scale of turbulence L_o is the scale size at which energy is fed into the turbulence field. Therefore, in order to compute the probability distribution of the correlation distance of phase fluctuations, we need information about the variability of the scale L_o. This is not obtainable from the data shown in Figure 4.4.

From the theoretical and experimental data of Janes and Thompson [23], the correlation distance of phase for a refractivity spectrum of the form $\Phi(K) \cong K^{-11/3}$ can be inferred. Based on these results an approximate curve for the effect of the phase fluctuations is shown in Figure 4.5. In order to overcome these atmospheric limitations, there are several methods at our disposal. One, obviously, involves selecting the right atmospheric conditions, when the adverse effects are at a minimum. Another involves a sidelooking radar system consisting of more than one antenna. If a synthetic aperture system of two antennas is arranged in such a way that the spacing between the antenna is longer than the correlation distance of the field, then the second antenna can be used to assess the extent of the atmospheric influences by checking the result of antenna number one at the same point in space but at a different point in time. In this manner the atmospheric effects can be reduced.

4.4.2 The Influence of Atmospheric Refractive Index Irregularities on the Sensititivy of an Absorption Spectrometer Measuring Atmospheric Gases

In Chapter 2, dealing with the basic principles in molecular spectroscopy, we noticed that the molecular structure reveals itself through well-defincd absorption lines. Thus, in order to measure the structure using spectroscopic methods, we shall either have to vary the frequency of the electromagnetic wave (the laser beam) that passes through the gas to be investigated, or we shall have to transmit simultaneously several waves having different frequencies.

Information about the gas concentration is obtained by measuring the relative amplitude of the wave that has been subjected to an absorption process. However, we have seen from Section 4.1 on atmospheric effects, that the atmospheric refractivity irregularities cause the intensity of the waves to vary, and furthermore, the variations at the two frequencies are not necessarily correlated. Determining then the gas concentration along the laser path by measuring the relative intensity at the two frequencies, a noise contribution stemming from the refractivity irregularities is added. This results in reduced system sensitivity.

An example illustrating this is given in Figure 4.6 [24]. This figure shows the relationship between spectral density of the fluctuations in the ratio of the two signals (k_1 and k_2) as a function of the fluctuation frequency. This spectral density is then a direct measure of the noise contribution from the atmosphere. Note that the smaller the ratio between the two wavenumbers k_1/k_2, the higher is the noise contribution. Note also that the solid line is the single wavelength log-amplitude spectrum.

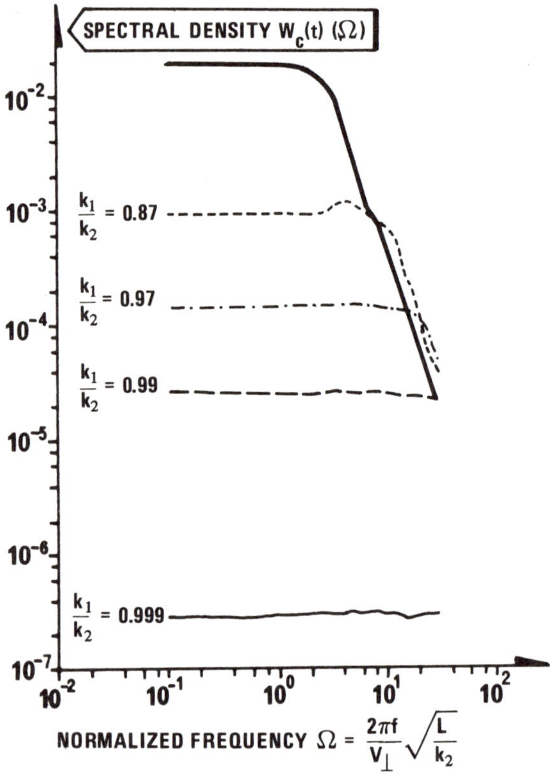

Figure 4.6 Theoretical spectral density resulting from refractivity irregularities for different wavelength ratios. The solid line is the single wavelength log-amplitude spectrum. After Kjelaas, Nordal and Bjerkestrand [24].

CHAPTER 5

REMOTE SENSING METHODOLOGY FOR THE DESIGN OF OPTIMUM SYSTEMS

In introducing the topic of generalized methodology [25], we should focus attention on two aspects. The first is concerned with the opportunities that new technology in the field of physics and electronics offers; the second is concerned with the task of matching new technological achievements to our particular problem of environmental surveillance by remote probing.

1. *Significant technological advances* during the last few years in the fields of microwave technique, electrooptics and computer technology open new avenues with regard to detection and identification of objects and of chemical agents with previously unidentifiable properties in our environment.

- *Wide-range tunable microwave sources* make it possible to illuminate the object under investigation over a frequency band in a predescribed manner for the particular target "fingerprint" to be revealed with optimum contrast and sensitivity.
- *Microwave receiver elements* will soon exist from which it is possible to determine the wavefront (amplitude and phase) resulting from scattering from an object under investigation over a large spatial area so that the geometrical shape of the object can be determined.
- Similarly, in the infrared and visible region, *tunable lasers* are available over a large range of wavelengths. Thus an illumination can be chosen so as to (1) penetrate the medium (the atmosphere) between the observer and the object of interest,

and (2) give maximum information about the molecular structure (the chemical compounds) of the object surface and of the agents emitted from the object (exhaust gases, oil spills, fragrance).

- *Electrooptical receiving systems,* specifically superheterodyne receivers, will soon, as was the case with microwaves 30 years ago, give a dramatic improvement in sensitivity (typically 4 - 6 orders of magnitude in the infrared region) and wavelength resolution. By means of such a receiving system and a coherent tunable source, the single-to-noise ratio of the signals scattered back from the object under investigation makes it possible to analyze the scattered (reflected) signal in great detail. Specific information about target molecular structure can be obtained.
- *Low-cost electronic data processors* (microprocessors) make it possible to perform the intricate and previously time-consuming computations necessary in real time (parallel processing).
- *Methods in electromagnetic field theory* have been developed from which the shape of the reflecting body (target in a background of interfering bodies) can be determined from measurements of the scattered EM field in space, time and frequency.
- Finally, *statistical estimation methods* such as Kalman filtering and pattern recognition methods have been developed. From these it is possible to make optimum use of a complex set of information (multisensor information about a target) so as to increase the overall sensitivity and the confidence level of the observations.

The preceding decade has been characterized by an endeavor to explore target-imaging methods over a large spectral range (side-looking microwave radars, optical and IR scanners). This is not likely to be the optimum approach when the task is to *detect, identify and provide quantitative information about* specific objects. The next decade is expected to be characterized by *dedicated* detection and recognition methods that make optimum use of available components and techniques.

2. *A fundamental advantage of significant potential value* in connection with the environmental surveillance problem is based on

the knowledge we already have about the object (or environmental parameter) of interest. The merits of this situation can be summarized as follows:

- The search (detection, identification, quantization) is very much simplified when we know that to look for. Knowing, for example, the detailed reflection or absorption spectrum of the chemical agent of interest (*e.g.*, lead deposit on vegetation from the burning of high octane fuel), the illumination employed can be structured so as to give enhanced detection capability for that particular target. In network theory this concept is known as "the matched filter."
- The identification may be based on several different properties of the object of interest. In this way a set of information can be obtained from which the confidence level of the information is significantly enhanced. The following example may illustrate this point. The object is to detect minerals rich in copper on the earth's surface or subsurface. The existence of such minerals gives rise to at least three independent sets of signatures, on the basis of which the mineral can be detected and identified:

  a. The soil rich in copper leads to copper being present in the vegetation growing on the particular soil. This can be detected remotely by emission or reflection spectroscopy.
  b. The composition of soil influences the species that will grow on it. Identification of the particular vegetation being present (or absent) will give us information about the soil and its composition. For example, the vegetation may be identified by measurement of its geometrical shape using wavelengths comparable to the geometrical scale (size) of the object of interest.
  c. The particular mineral may under certain conditions contribute to the composition of the air above the ground. For example, measuring the amount of sulfur compounds in the air above the ground by laser spectroscopy methods gives indirect information about the surface composition.

  The degree to which a given parameter (signature) is dominant depends on the prevailing conditions. By a suitable weighting procedure (*e.g.*, Kalman filtering process) the information set can be processed so as to uncover previously unidentifiable properties and provide unambiguous classification.

The following example may substantiate the arguments above. The objective is to detect a vehicle in a background of loose rocks. You know the shape, size and chemical surface structure of the target of interest. Hence you know *what* to look for. You are seeking two bits of information only: is the target there or it is not? From an information point of view only, an infinitesimally small bandwidth is required to make this assessment. For a given "signal power" (radio power scattered from the object of interest) the signal-to-noise ratio approaches infinity as the information bandwidth of the signal approaches zero. Any signal, however small, can in principle be extracted from a background of noise, however large, if the signal is a pure sine wave of known frequency and sufficient integration time is provided.

We shall now proceed to elucidate the statements made above. The objective of this chapter is to provide the basis for an optimum design of specific remote sensing systems. The design philosophy is based on the following, which we have to our disposal:

1. A set of sensors (see above).
2. Knowledge about the basic principles on which the action of these sensors is based, knowledge about the interaction between the electromagnetic waves to which these sensors are coupled and the environmental factor that we seek information about (Chapters 2 and 3).
3. Information about the limitations imposed by the propagation medium between the observation platform and the object of interest.
4. Knowledge about the shape, size and chemical composition of the object of interest (signature information), and, in some cases, about other possible objects (interferents).

We shall be concerned with the design of remote sensing systems, which in an optimum way make use of available knowledge and of available sensors. We seek a system where the illumination, *i.e.*, the electromagnetic waves, are matched to the parameter to be measured in an optimum manner in an environment of "noise" (matched illumination), and we seek a set of detecting sensors which are matched to the waves that are reflected, emitted or absorbed by the object of interest (matched reception) [25].

5.1 STATEMENT OF THE PROBLEM

In brief, the problem can be described as follows. A method of optimum sensitivity for the detection, identification and quantitative evaluation of an object or a chemical agent is sought. It is assumed that all the pertinent data pertaining to the target are known. These *target "fingerprints,"* are typically:

- The shape, size and texture of the object, *i.e.*, the distribution of scattering elements in height, width and depth (the macrostructure).
- The chemical composition of the surface or surface layers.
- The chemical composition of the gases (or liquids) that the object of interest may emit (scent from plants, odors from fermenting organic substances, gases from soil, exhaust gases from vehicles).
- The footprints, tracks or traces that a moving object leaves behind it.

In general, the target will possess useful fingerprints in various spectral ranges. Some of these are weak and some are pronounced. When it comes to determining which of these fingerprints to utilize, we shall also have to consider the *fingerprints of the background,* the noise, and the properties of the medium through which the illuminating waves must pass in order to illuminate the objects. Thus, consideration will have to be given to the following:

- Characteristics (fingerprints) of the target (deterministic or statistical).
- Characteristics (statistics) of the background (additive noise).
- Characteristics (structure) of the propagation medium between the platform of observation and the target (distortion, attenuation, multiplicative and additive noise).

Since at least the two latter factors vary in time and space, it is not likely that there exists a unique frequency band within which optimum detection and identification capability is achieved at all times. Therefore, one would wish to have observations in multiple fingerprint regions and sort out the best set of data under any given condition.

We shall now proceed to discuss the two main aspects of the general complex of problems briefly mentioned above. These are:

1. Detection and identification based on chemical properties of the object surface.
2. Detection and identification based on object shape.

5.2 OPTIMUM DETECTION AND IDENTIFICATION OF A CHEMICAL AGENT (CHEMICAL COMPOSITION OF TARGET SURFACE)

In order to illustrate the essential points and the potentials of the technique, let us define a challenging problem of practical importance, using it as an example. Our aim is to detect and identify a particular chemical agent, for example some sulfate deposited on vegetation in a certain geographical area. This could be an area contaminated from human activity or by deposits from general air pollution. *A priori* this agent may have its fingerprint (absorption lines) in any wavelength region from the ultraviolet region (λ = 0.2-0.4 μm) through the infrared region (2-20 μm) to the millimeter and microwave region (λ = 100-200 μm).

To reveal these fingerprints, we illuminate the ground on which this agent may be present with electromagnetic waves, changing the frequency over the frequency band of interest in some prescribed manner that is optimum with regard to the detection and identification of the particular chemical compound of interest. The situation is assumed to be as follows:

- We are looking solely for a specific chemical compound.
- We know the absorption spectrum or reflectance spectrum of this compound.
- We know nothing about any of the other agents (interferents) that may be present.
- We have very meager information about the vegetation (topography) on which the agents may be deposited, but we have some general idea about the roughness of the structure (grass, coniferous trees, rocky ground). Let us assume that the background is coniferous trees.

A hypothetical detection system will be discussed for the purpose of illustrating the essential points. Figure 5.1, which is subdivided into two sections, shows a remote probing system in symbolic form. Section A has a generator G providing the illumination. It illuminates an area on which contaminated trees are growing. The contaminating agent is assumed to be deposited over

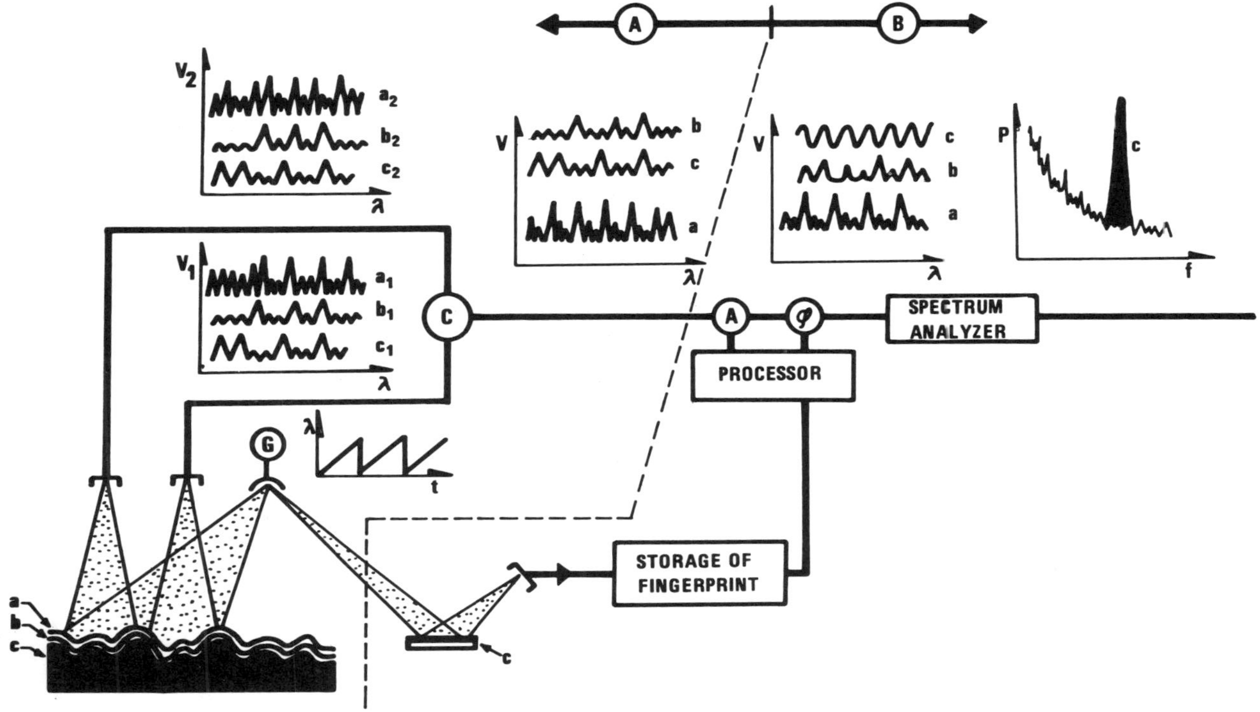

Figure 5.1 Symbolic diagram illustrating the optimum detection and identification principle.

the area, which is viewed by two receivers having two separate not overlapping fields of view. When the frequency of the illuminator is changed in a linear manner (saw-tooth frequency-modulation) and the signals V_1 and V_2 , which are received at the two receivers, are observed, (referring to the symbolic diagram of Figure 5.1), the signals have the following components:

- A random signal with large variance (Rayleigh distributed) resulting from the topography. The signal is a result of backscatter from scattering facets distributed in depth. They give rise to many interfering waves and result in large variations in field strength with frequency (signal trace marked "a").
- A deterministic signal component resulting from the complex surface chemistry (trace marked "b") appearing as modulation of the random topography signal.
- Deeply imbedded in the background chemistry is the component of interest, namely that stemming from the absorption spectrum of the particular chemical compound of interest (trace marked "c").

Our first task in the signal enrichment process is to reduce the dominating effect of the topography (traces a_1 and a_2). In order to achieve this, we shall make use of knowledge we already have about the signal (the deterministic molecular absorption spectrum) and about the noise (the random signal resulting from scattering from a random surface such as from a tree). The noise resulting from the scattering process will obviously be statistical in nature, and there will be no correlation between the signals at the two receivers if the vegetation within the two separate fields of view is not identical in form. Furthermore, if the scatterers (the branches of the trees) move only a fraction of a wavelength from one frequency scan to the next, the correlation from one frame to another at either receiver will be limited. Since the turbulence scale (*i.e.*, the spatial correlation distance) of the surface wind causing the vegetation to move is small compared with the height of the vegetation, the above assumptions are justified.

Therefore our task is to find a process by which the uncorrelated noise component can be reduced relative to the correlated signal component. This is done in the correlator marked C in Figure 5.1. To ensure a thorough physical understanding of the statistical signal retrieval methods involved, this process will be discussed in some detail.

First let us consider the case where the two contributions are added together (additive noise). Figure 5.2 gives a plot in three dimensions of the data that form the basis for the correlation process. Consider the signals from receiver No. 1 first. In our cartesian coordinate system the signal strength is plotted vertically,

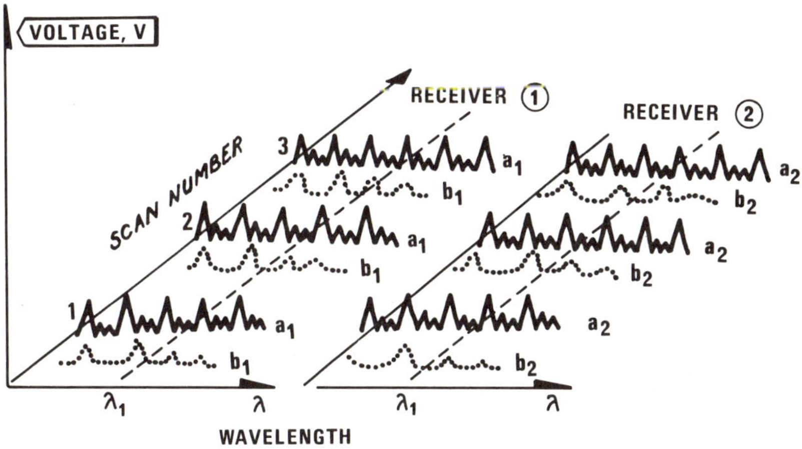

Figure 5.2 Data forming the basis for signal enhancement by a correlation process.

wavelength is plotted horizontally in the plane of the paper and the frame-number (or scan number) is plotted orthogonally to the V-λ plane. The signal resulting from the deterministic absorption spectrum is denoted "b" as in Figure 5.1 and the random signal stemming from the vegetation is denoted "a." To reduce the effect of the random uncorrelated contribution, we form an ensemble covariance function for each value of λ, by comparing (correlating) $V(\lambda_1)$ from each scan. Thus we will obtain a number of points on the covariance function determined by the number of frames (scans) and we will obtain a number of covariance functions determined by the wavelength resolution Δλ. If our ensemble then is N, our signal to noise ratio is enhanced by a factor $\sqrt{N}$.

However, the simple example with additive noise (addition of coherent and noncoherent components of the signal) is probably not the most realistic one. For example, if the coherent signal appears as an amplitude modulation on the incoherent signal, then it would be more meaningful to cross-correlate the spectral components of the signals from the two receivers by forming the coherence function instead of the covariance in the C processor.

In order to express this mathematically, we will use an approach based on work by Kjelaas, Nordal and Bjerkestrand [24]. In the general case we may have both multiplicative and additive noise. If the spectral components of the noise are not overlapping those of the signal, a simple filtering technique can be employed. If the spectral components do overlap, we shall have to use methods that give us an optimum signal-to-noise ratio. First let us consider the case of *additive noise.*

If then S(t) be the signal and n(t) the noise, the resultant

$$F(t) = S(t) + n(t) \tag{5.1}$$

We then assume that there is no correlation between S(t) and n(t). The cross-covariance R_{fs} between the resultant signal f and the signal s is given by

$$R_{fs} = R_{(s+n)s}(t) = R_s(t) + R_{ns}(t) = R_s(t) \tag{5.2}$$

Similarly the autocovariance is

;R

$$R_f(t) = R_{s+n}(t) = R_s(t) + R_n(t) \tag{5.3}$$

This can be written in the form

$$[R_s(t) + R_n(t)] \otimes h(t) = R_f(t) \tag{5.4}$$

where the symbol $\otimes$ denotes convolution and h(t) is therefore the impluse response. The transfer function $H(\omega)$, which is the Fourier transform of the h(t) function, is then given by

$$H(\omega) = \frac{P_s(\omega)}{P_s(\omega) + P_n(\omega)} = \frac{P_s(\omega)}{P_f(\omega)} \tag{5.5}$$

where $P(\omega)$ is the power spectrum so that $P(\omega)$ is the Fourier transform of the autocorrelation function R(t).

Let us consider the case of a multiplicative noise source. The resultant signal may be written as

$$f(t) = S(t) \otimes h_n(t) \tag{5.6}$$

and

$$R_f(t) \otimes h_n(t) = R_{fs}(t) \tag{5.7}$$

which in terms of the transfer function gives

$$H(\omega) = \frac{R_{fs}(\omega)}{R_f(\omega)} \tag{5.8}$$

We can now express the transfer function $H(\omega)$ in terms of the coherence

$$\text{coh}_{sf}(\omega) = \frac{P_{sf}(\omega)}{P_s(\omega)\, P_f(\omega)} \tag{5.9}$$

where $P_{fs}(\omega)$ is the cross-spectrum and the transfer function is given by

$$\begin{aligned} H(\omega) &= \frac{P_{sf}(\omega)}{P_f(\omega)} \\ &= \frac{\text{coh}_{sf}(\omega)\, P_s(\omega)\, P_f(\omega)}{P_f(\omega)} \\ &= P_s(\omega)\, \text{coh}_{sf}(\omega) \end{aligned} \tag{5.10}$$

Thus, if we are dealing with additive uncorrelated noise, the signal-to-noise ratio is enhanced by forming the covariance function as implied in the discussion of Figure 5.2 above. The operator C in Figure 5.1 is a correlator giving us the ensemble covariance function. If, however, the noise is multiplicative, we shall have to cross-correlate the spectral components by forming the coherence function.

Concluding the discussion of section A in the symbolic Figure 5.1, we note that the uncorrelated effects of the rough scattering surface are suppressed during the correlation process. All the correlated factors, chemical compound of interest deeply imbedded in all the uninteresting molecular structures, are retained in unaltered proportions. Now let us proceed to section B of the diagram.

This section illustrates a method by which we can make use of the detailed information that we possess about the chemical compound of interest (molecular structure, reflectance spectrum) to perform a selective detection.

- The signal preprocessed in section A is "filtered" with the fingerprint (reflectance spectrum) of the chemical compound of interest. In the symbolic diagram of Figure 5.1 this is achieved by shining part of the energy from the illuminator onto a reflector containing a film of the chemical agent on which we are focusing our attention. The resulting reflectance spectrum contains the fingerprints that give us the basis for the second filtering operation.

- The operation involves a structuring of the original frequency sweep so as to limit the frequency content of the signal resulting from the chemical agent of interest to a delta function in the frequency domain (*i.e.*, to a pure sinusoidal variation in time over the sweep period). In a practical system one would accomplish this by sweeping the transmitter frequency (the illuminator) in a particular manner (not with a saw-tooth as depicted in the schematic figure above) and at the same time and in synchronism with the frequency sweeping, amplitude modulate the illuminator (see Figures 5.3 and 5.4).

To illustrate this principle and its merits, let us assume that the ϕ and A operators of Figure 5.1 is a tape recorder. The signal is played into the tape recorder in real time and played off successively. The first operator, the "A function," involves an amplitude modulation of the signal so as to adjust the lines in the reflectance spectrum related to the agent of interest to the same level. This involves an amplitude modulation of the reflectance spectrum of interest. The second signal from the processor (the ϕ operator) adjusts the speed of the tape to obtain constant spacing between the lines in the reflectance spectrum of interest. If the interval in the wavelength domain is large between two lines, the tape should be speeded up; conversely, if the distance is small the speed should be slowed. In this manner all the characteristic spectral lines will be equally spaced when originating from the chemical compound of interest and distributed in a disordered manner for all the interferents.

In principle, adjusting both the frequency and the amplitude of the generator illuminating the target of interest in great detail, can transform the signal component originating from our particular chemical compound to a pure sine wave requiring a scan-time limited bandwidth to be detected. In this manner the signal-to-noise level of our system, can, in principle, be brought to a very high level. Limiting factors are determined by the available

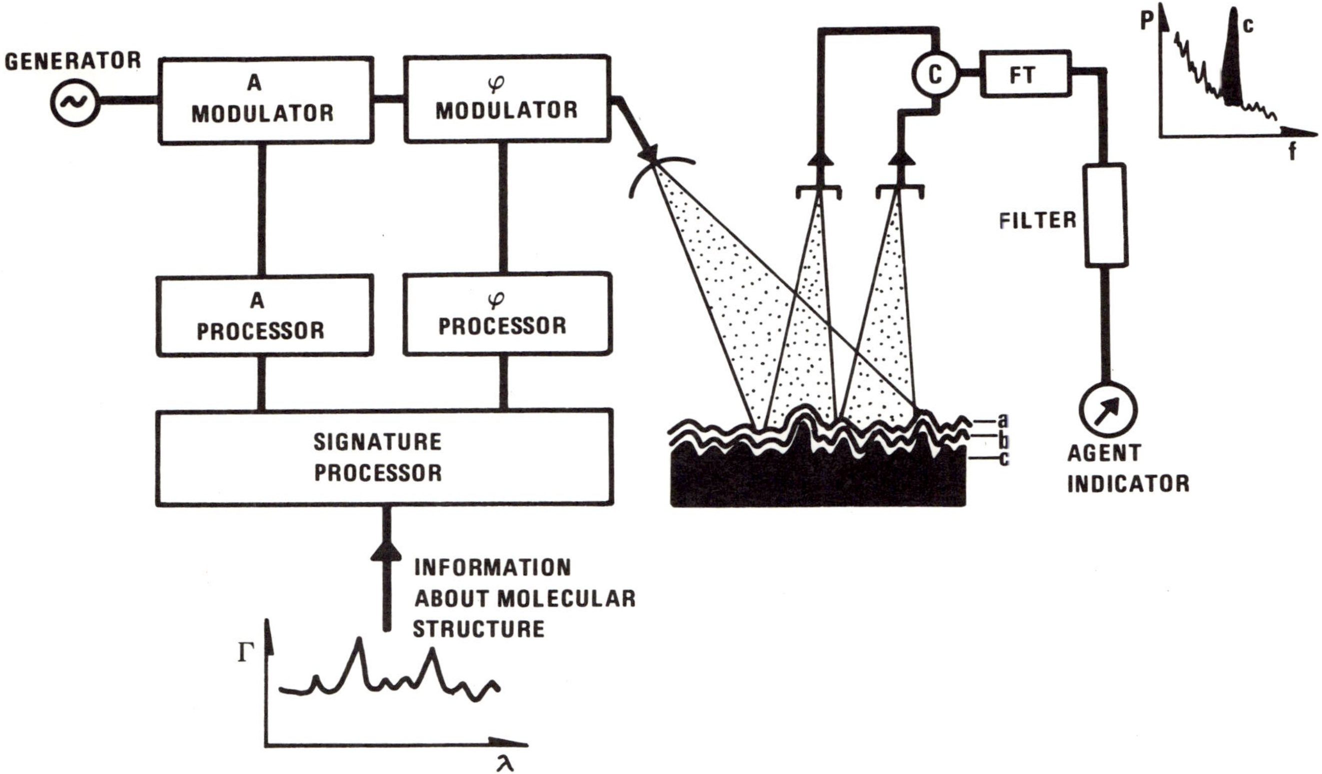

Figure 5.3 Schematic diagram of a practical system with matched illumination.

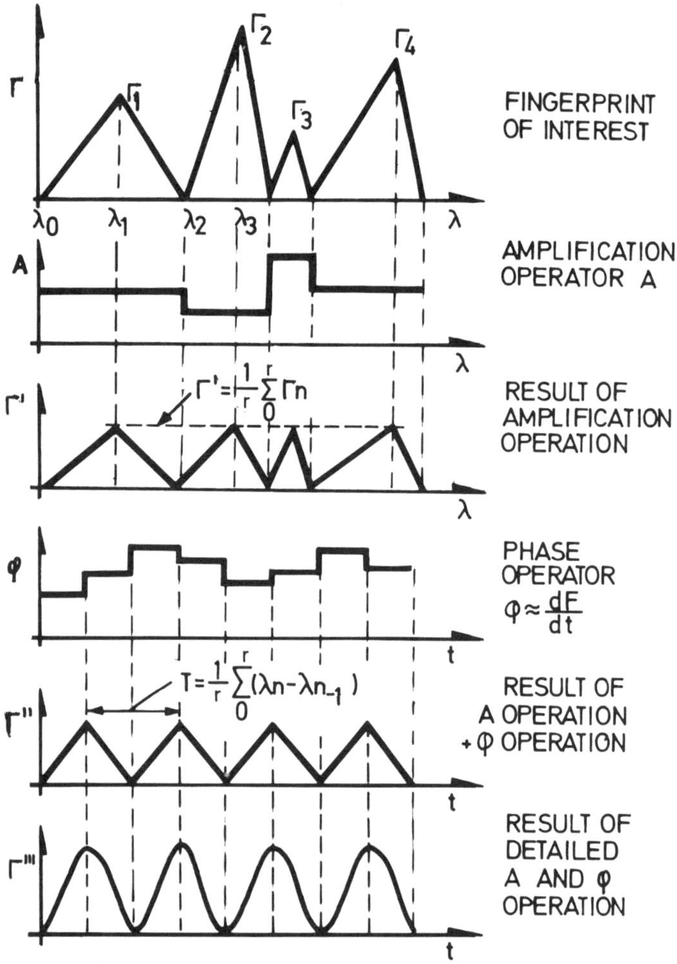

Figure 5.4 Example of an optimum structuring of the illumination function for the purpose of detecting/identifying a particular molecular structure.

integration time and the degree to which it is practicable to structure the illumination function.

Finally, the basic Figure 5.1 shows the result of a final Fourier analysis of the signal. Our signal of interest is concentrated to a very narrow spectral region, whereas the effect of the interferents is distributed over the spectrum. Before we proceed to consider in some greater detail the character of the noise to be suppressed, we shall give a more realistic example of a selective detection method. Such a method is illustrated schematically in Figure 5.3.

The illumination (microwaves, IR, visible light, UV) is amplitude and frequency modulated by particular waveforms (see Figure 5.4). These waveforms are the result of a detailed processing of the molecular signatures (see Figure 5.4). The processed illumination is then transmitted to the object of interest. Since the illumination is matched to the molecular structure of interest, the signal appearing at the receivers has minimum information bandwidth and the bandwidth of the entire receiving system can be minimized. Minimum bandwidth gives minimum noise contribution.

Figure 5.4 illustrates the A and ϕ operation process. In the top curve of Figure 5.4 an idealized molecular spectrum is shown (spectrum of absorption, emission or reflectance). The reflectance Γ varies in a triangular manner with wavelength. If the illuminator is linearly frequency modulated, this is what the signal would look like. Amplitude modulating the illuminator in a manner illustrated in Figure 5.4, second curve from above, seeking the same strength of all the lines in the reflectance spectrum, gives the results shown.

Referring now to the fourth curve from above, we change the relative position of the maxima and of the minima of the spectrum so as to obtain a periodic function shown in the fifth graphical representation. This is achieved by changing the rate dF/dt at which the frequency is changed. It is then obvious that if a more detailed A and ϕ operation is applied, the result is a sinusoidal variation, shown at the bottom of Figure 5.4. To detect this requires only a very small bandwidth.

For the purpose of emphasizing the merits of the technique, some illustrative examples are given in Figure 5.5. Two different absorption spectra are considered. Type "A" is characterized by eight absorption lines, whereas type "B" has ten absorption lines. Each absorption line is Lorentz shaped. In the interval between the lines the absorption/reflection is assumed to be zero.

Adopting the amplitude- and frequency- modulation scheme illustrated in Figure 5.4 above, the illumination is structured for type "A" molecules. Figure 5.5 shows the result of this optimization process. Note that the spectrum density function associated

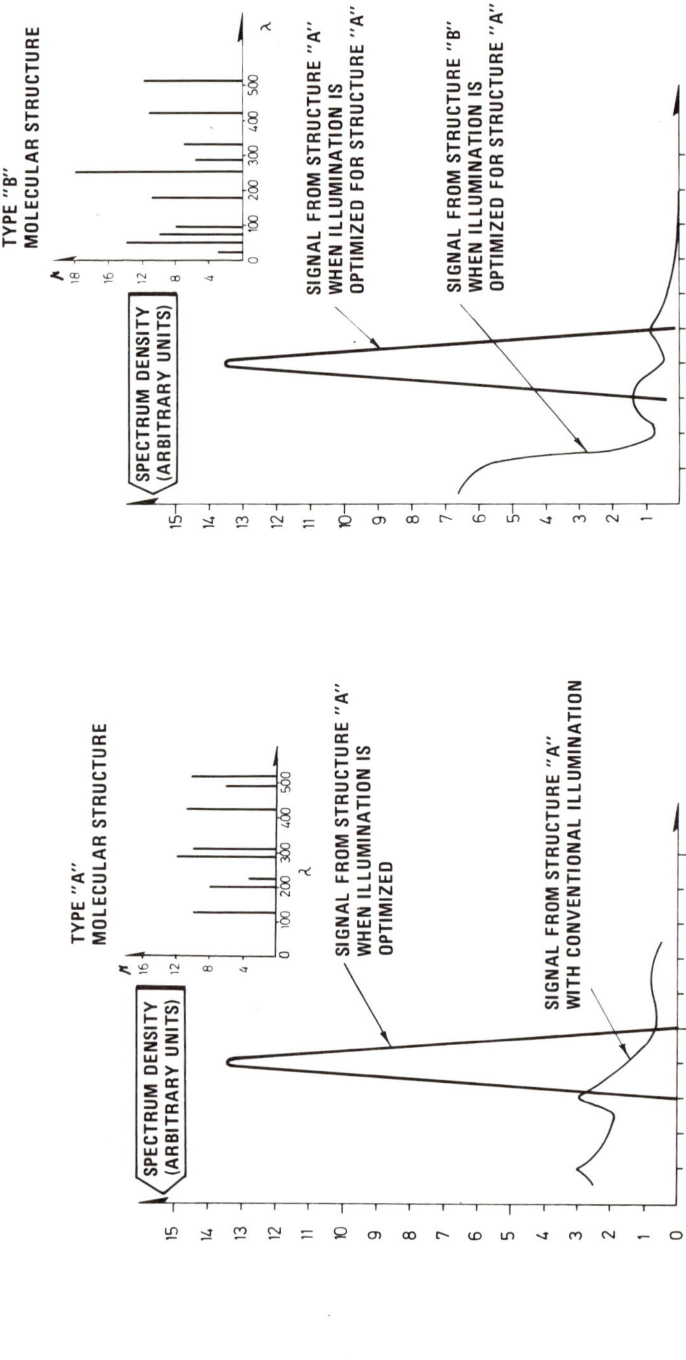

Figure 5.5 The effect of optimized illumination; theoretical results.

with optimized illumination is narrow, whereas the spectrum resulting from conventional illumination is wide. Also, if an extensive structuring of the illumination had been accomplished, the resulting signal spectrum would have been a delta function. Figure 5.5 also shows the discriminating power of the technique (right-hand figure). Here an illumination structured for type "A" molecules is applied to structure "B." Note the marked difference in maximum spectral intensity and the shift in frequency of the peaks.

If a computer simulation program implementing this crude technique that involves "first-order matching" only for three frequently encountered compounds, namely epoxy, vinyl and biphinyl, is used, the result is Figure 5.6. Note that by the use of narrow filters as depicted in Figure 5.3 above, the particular agent of interest can be selected when its absorption spectrum is known.

Let us now express this mathematically. On the basis of information about the absorption spectrum, we search the expression for the manner in which the frequency should be varied so as to give a sinusoidal variation of the received (reflected) signal. This we refer to as matched illumination.

For a linear frequency scan, the frequency as a function of time is given by

$$\omega = \omega_1 t' + \omega_0 \tag{5.11}$$

where ω_1 and ω_0 are constants.

If $f(\omega)$ is the reflection/absorption spectrum, applying linear scanning we have

$$f(\omega) = f(\omega_1 t' + \omega_0) \tag{5.12}$$

We now want a modified sweep function giving us a periodic output function. Hence, we wish to obtain

$$f(\omega_1 t' + \omega_0) = \hat{f}(t') = g(t) \tag{5.13}$$

where g(t) is our periodic function. This can be achieved by inverting the f(t') function. Thus, the function should satisfy the condition

$$f^{-1}\,\hat{f}(t') = t' = f^{-1}[g(t)] \tag{5.14}$$

By substitution this gives

$$\omega(t) = \omega_1 f^{-1}[g(t)] + \omega_0 \tag{5.15}$$

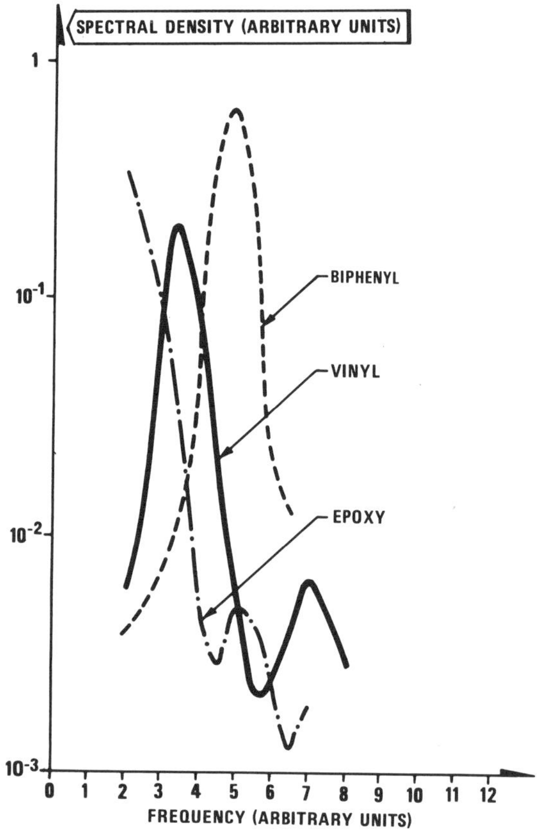

Figure 5.6 The effect of matched illumination for epoxy, vinyl and biphinyl. The illumination that has been structured for vinyl is applied to epoxy. Similarly, the curve for biphinyl is obtained by structuring the illumination for biphinyl. Computer simulations.

Therefore, in order to match the illumination to the agent of interest, a frequency sweep function $f^{-1}(\omega)$, which is the inverse of the reflection/absorption spectrum $f(\omega)$, should be used.

An example will clarify this. Let the absorption spectrum be of an exponential form

$$f(\omega) = e^{-a\omega} = g(t)$$

$$f^{-1} f(\omega) \rightarrow \omega = -\frac{1}{a} \ln g(t) \qquad (5.16)$$

$$\omega(t) = -\omega_1 \frac{1}{a} \ln g(t) + \omega_0 \qquad (5.17)$$

Having given the general expression for the matched illumination function in terms of functions describing the material of interest, we shall go on to consider the effect of the topography (the surface roughness) in relation to that of surface chemistry. The question we now ask is: how should the transmitting/receiving system be structured in order to give maximum information about the geometrical shape of an object? Or, how should the system be structured to have optimum sensitivity in relation to a given object of known size and shape.

5.3 OPTIMUM DETECTION AND IDENTIFICATION OF THE BASIS OF GEOMETRICAL SHAPE OF OBJECT

When making use of the geometrical shape of the object for detection purposes, we shall, contrary to the above, choose frequencies that are not affected by the detailed molecular structure of the scattering surface. In the absence of this effect, there are three remaining effects that are of importance:

- the multiplicative effect of atmospheric irregularities (turbulence) and of particular matter (rain, snow, dust)
- the additive effect of the background scatterers.

The first effect determines the frequency band to be employed. The higher the frequency of the illuminator, the more severe is the impact of atmospheric irregularities. The shape of the background elements relative to that of the target determines the detailed structuring of the illumination function. To establish an illumination function there are four degrees of freedom, space and time or frequency. The illuminating wave can be structured in the frequency domain, and the wave can be structured in space (shape the wavefront) so as to match the target of interest. First let us consider the frequency domain.

5.3.1 Optimum Frequency Distribution of an Illuminating Wave with Regard to a Target of Known Geometrical Form

In this section we shall investigate the information obtainable by modulating the illumination of the target in time. There are, as we well know, several ways of doing this. We will discuss *one* powerful method.

We have shown in Section 3.3 that, knowing the distribution in depth of the scatterers, we can calculate the correlation properties in the frequency domain (bandwidth) of the signal reflected (scattered) back from the target [26]. We shall use the results of this reference for the purpose of establishing the optimum *frequency distribution* of the illuminating wave to determine the distribution in depth (longitudinal) of the scatterers constituting the target.

To illustrate the principle employed, two figures are presented. In Figure 5.7 a target with a pyramidal shape is illuminated by a set of waves whose frequency spectrum is rectangular, *i.e.*, the

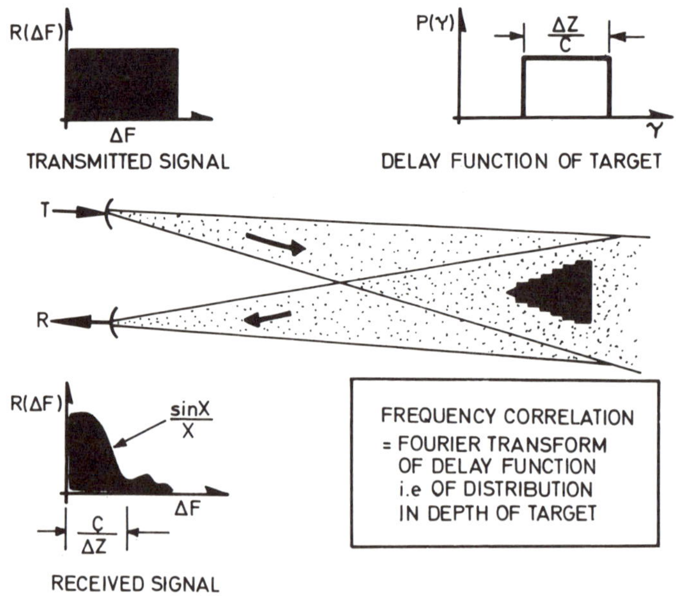

Figure 5.7 The basic principle–distribution in depth of target.

correlation properties of the illuminator in the frequency domain is a rectangular function. The delay spectrum of the pyramid, which has a rough surface in order to give diffuse reflection, is also a rectangular function. We have shown (Section 3.3 and Reference 25) that the effect of a target having a finite depth is to decorrelate the frequency properties of the waves scattered back from the target. Specifically, if a set of waves correlated in amplitude and phase impinges on a target having a given distribution in depth of the scatterers P(z/c), then the correlation properties of the reflected waves are limited to some function of the Fourier transform of P(z/c). Accordingly, as depicted in Figure 5.7, the transmitted rectangular correlation function is reduced to a (sin x)/x function upon reception if the target delay spectrum is rectangular.

The identification potential of this multifrequency system is illustrated in Figures 5.8 and 5.9. Here the frequency properties of the scattered signal are calculated for a set of scattering objects having different shape and size. Note the very marked influence of object shape and the dramatic size dependence. Note also that the peak of the w(f) spectrum in Figure 5.9 is determined by the spacing of the branches, whereas the width of the curve is determined by the height of the tree. Table 5.1 gives the mathematical expression for the bandwidth. Note that there is a simple and direct relationship between bandwidth and correlation distance in the frequency domain (see Section 3.3).

It should be noted that Figures 5.7 and 5.8 together with Table 5.1 refer to the monostatic backscatter situation where both the transmitter (the illuminator) and the receiver are located at the same spot. In practice this causes some difficulties since we shall have to ensure a good isolation between the transmitting and the receiving antenna in order to prevent direct transmission from one to the other [26]. To overcome this, a bistatic arrangement may be used. This is illustrated in Figure 5.10.

From simple geometrical considerations, with reference to Figure 5.10 we find that the difference in path length $\Delta \ell$ can be expressed in terms of depth z as follows:

$$\Delta \ell \cong 2(\alpha + \frac{z}{d})$$

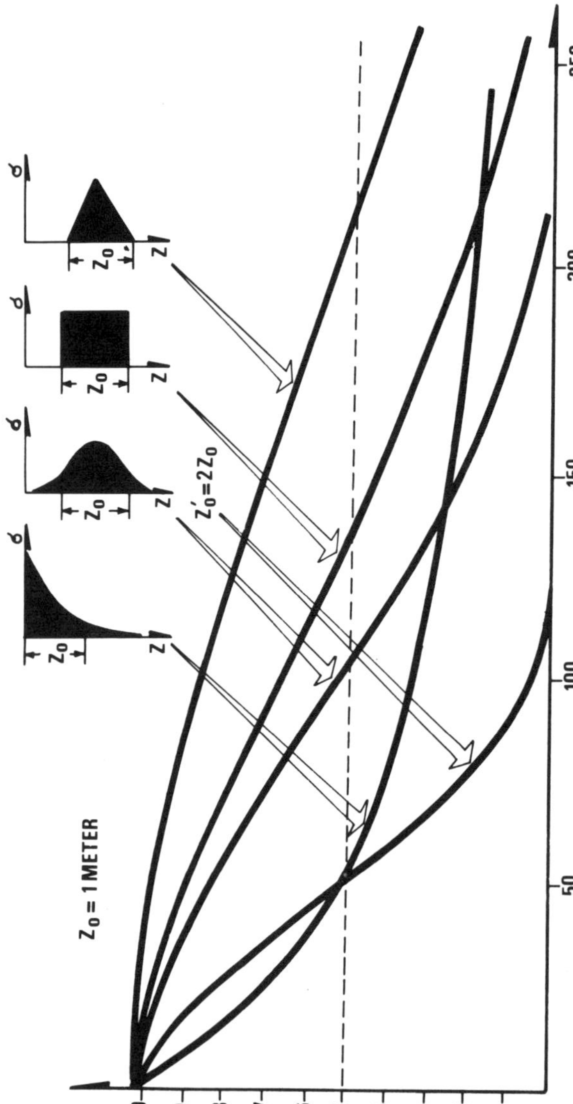

Figure 5.8 Bandwidth properties of some topographic surfaces. The thickness of the reflecting strata is taken as 0.5 m in the examples, so that z_o becomes 1 m.

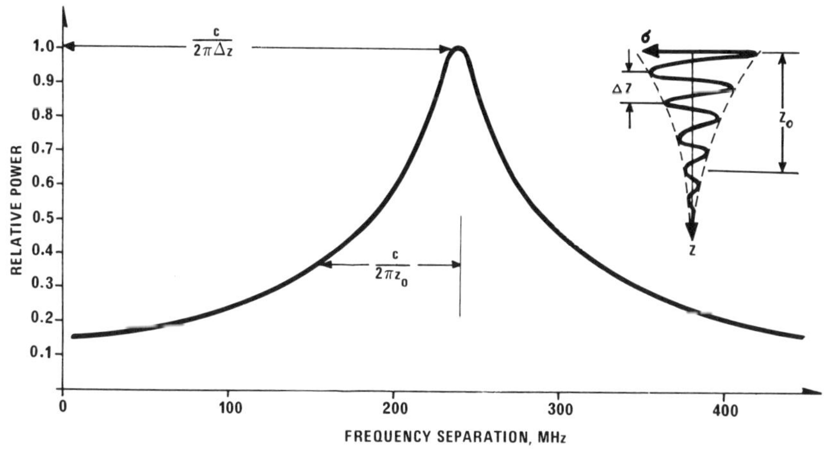

Figure 5.9 Bandwidth properties of a scattering surface in which there is a periodic structure, such as evenly spaced branches of a tree. In the example z_0 is 1 m and Δz is 20 cm.

where α is the scattering half-angle and d is the distance between transmitter and receiver as indicated in Figure 5.10. Thus, if the characteristic length of distribution in depth of the scatterers is z_0, and if the path length is 50 km and the angle α is 500/50,000 (implying that the vertical distance from the line between transmitter and receiver to the scattering area is 500 m), then the resulting delay $\Delta \ell$ is approximately $z_0/100$. The bandwidth of this reflector configuration is thus increased by a factor 100 relative to the pure backscatter case.

If, however, some of the transmitted power reached the receiver directly through the line of sight path, then the spread in path length $\Delta \ell$ is given by

$$\Delta \ell = \frac{2H_1 H_2}{d}$$

where, as before, d is the path length. H_1 and H_2 are the height above the reflecting surface of the transmitter and receiver respectively. In the example considered above $H_1 = H_2 = 500$ m and d is 50 km.

Under such multipath conditions the difference in path length is some 10 m, and the bandwidth is reduced by a factor 1000

Table 5.1 Bandwidth Properties of Some Topographic Surfaces

Distribution in Depth of Scatterers	Frequency Dependence of Reflecting Surface	Half-Power Bandwidth
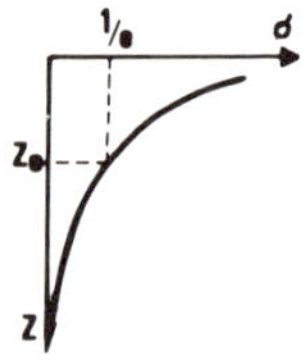 EXPONENTIAL	$W(\omega) = \frac{\sigma_o^2}{(\frac{c}{z_o})^2 + \omega^2}$	$\Delta F_{1/2} = 0.16 \frac{c}{z_o}$ Hz
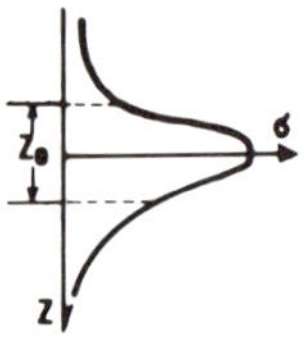 GAUSSIAN	$W(\omega) = \frac{\pi}{4} \frac{\sigma_o^2}{(\frac{c}{z_o})^2} e^{-\frac{\omega^2 z_o^2}{8 c^2}}$	$\Delta F_{1/2} = 0.37 \frac{c}{z_o}$ Hz
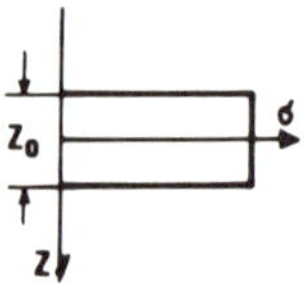 RECTANGULAR	$W(\omega) = 4\sigma_o^2 \left(\frac{\sin \frac{z_o}{2c}\omega}{\omega}\right)^2$	$\Delta F_{1/2} = 0.44 \frac{c}{z_o}$ Hz
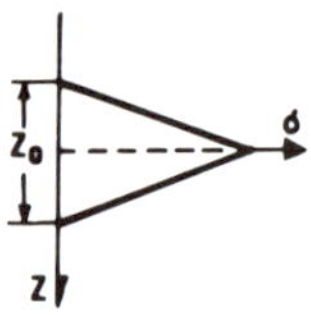 TRIANGULAR	$W(\omega) = \frac{64\sigma_o^2}{(\frac{z_o}{c})^2} \left(\frac{\sin \frac{z_o}{4c}\omega}{\omega}\right)^4$	$\Delta F_{1/2} = 0.64 \frac{c}{z_o}$ Hz
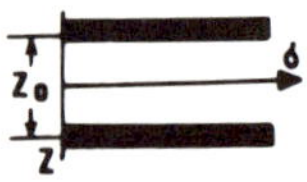	$W(\omega) = 4\sigma_o^2 (\cos \frac{z_o}{2c} \omega)^2$	$\Delta F_{1/2} = 0.25 \frac{c}{z_o}$ Hz

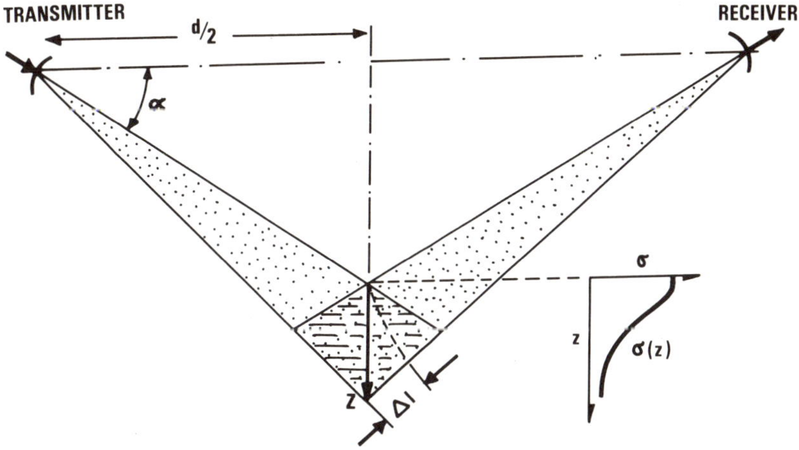

Figure 5.10 The geometry of the forward scatter situation.

relative to the case discussed above with forward scatter and no direct line of sight path. Note that the locus of the reflection point for constant delay is an ellipsoidal surface with the transmitting and the receiving terminals as focal points. For the present geometry, therefore, the constant delay surfaces are horizontal and parallel near the midpath point. As a consequence of this, it is the distribution of the scatterers [*i.e.*, the $\sigma(z)$ function] along a vertical direction in space normal to the constant delay surface that matters. Note also that the current treatment assumes that the scattering surface is sufficiently smooth for shadowing effects to be neglected, as is focusing due to large curved surfaces. For details on this subject, the reader is referred to [14] and [15].

In conclusion, when the correlation properties in the frequency domain of the signal scattered back from the target are measured, information about its distribution in depth is obtained by Fourier transformation of the received signal. This simple, direct method would have been effective if the target were suspended in the air with no interfering background. In the more realistic case of an interfering background, a *selective detection method* must be used. The procedure is analogous to that discussed in some detail in connection with the study of molecular structures. We should match the illuminating wave to the target of interest so as to obtain a narrow band signal of known frequency back from the target of interest, and a signal distributed in frequency from all the interferents, the background. Figure 5.11 illustrates the principle of operation.

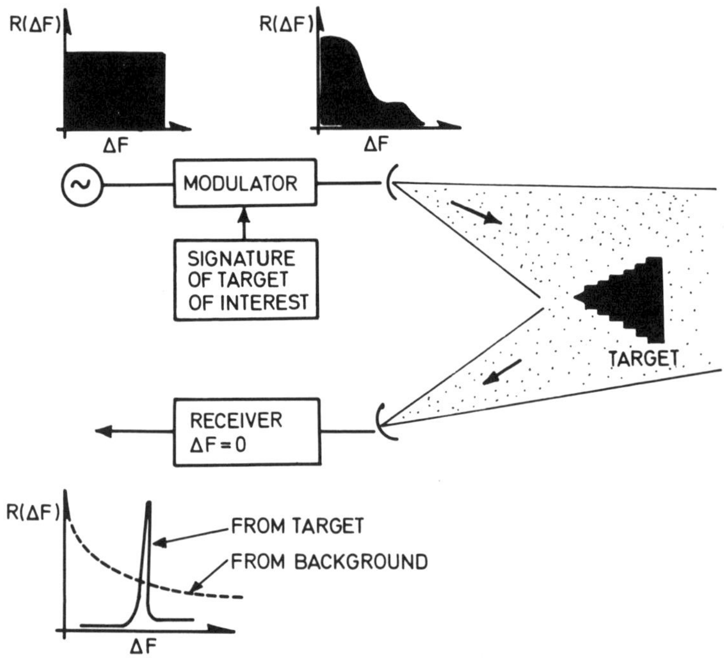

Figure 5.11 Illumination function for optimum detection sensitivity. Our system has optimum sensitivity for objects of interest and minimum sensitivity for other objects (background).

On the basis of the principles discussed in Section 5.2 above, we have structured the correlation properties of the illuminating waves so as to match them to the object of interest (see Figure 5.12). In our example, the illumination is matched to an object 50 cm deep having a Gaussian distribution of the scattering surface. This gives rise to a narrow-band signal from the object of interest, whereas the returns from targets to which the illumination is not optimized are relatively broadbanded. Note that there is no unique relationship between the delay function $\sigma(z/c)$ and the shape of the object. Obviously, the variation in z has to be restricted to the boundary of the object, but there are a large number of geometrical shapes that give the same delay function.

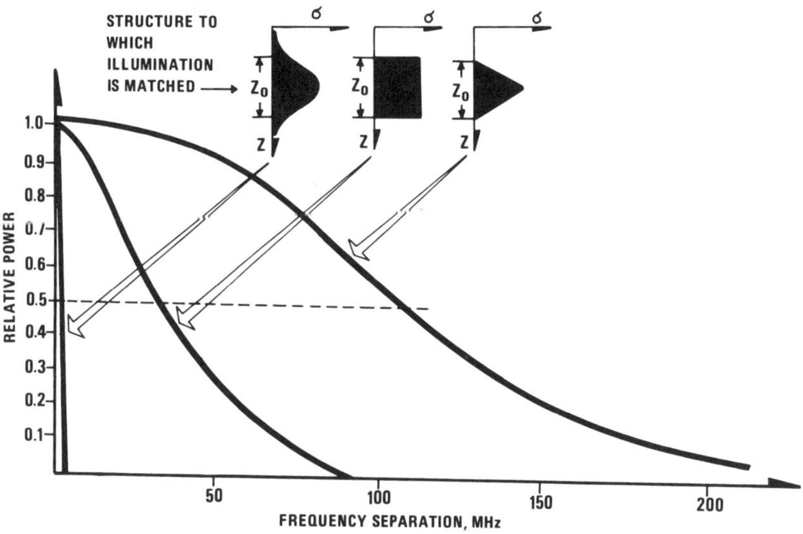

Figure 5.12 Theoretical results of matched illumination. The illumination is matched to the "Gaussian object."

Finally, let us investigate the information obtainable about an object by measuring the spatial correlation properties of the waves scattered from an object of a given shape.

5.3.2 Determination of Target Distribution in Space by Spatial Correlation Measurements of Field Strength

The above discussion dealt with an experimental determination of the distribution in depth (in the direction of propagation) of the scattering elements constituting a scattering object. If we illuminate an object by a set of EM waves with mutually correlated amplitudes and phases, this set of correlated waves is subjected to a decorrelation process. The degree to which the waves are decorrelated depends on the distribution in depth of the scattering elements constituting the object. Specifically, the correlation function in the frequency domain (covariance function for waves having separate frequencies) is the Fourier transform of the distribution in depth of the scatterers (Fourier transform of delay spectrum).

We will now focus our attention on the spatial correlation properties of a scattered wave in relation to the spatial distribution of scatterers over the object. We have shown in Section 3.2 that in exactly the same way as for spaced frequencies the transverse (normal to direction of propagation) spatial correlation of the complex field strength is the Fourier transform of the angle of arrival spectrum *i.e.,* the Fourier transform of some function determined by the transverse distribution of the scatterers constituting the target of interest. Thus, if X be a distance coordinate orthogonal to the Z-direction of propagation, and if $\sigma(X)$ be the distribution of scattering elements over the scattering object under investigation, the spatial correlation of field strength in the X-direction is given by

$$R_E(x) = \text{Fourier transform of } \sigma(X/R)$$

R being the distance between the point of observation and the target. Note that $\sigma(X/R)$ is the angular power spectrum of the scattered wave. Figure 5.13 illustrates the basic principle related to the determination of angular power spectrum on the basis of field strength correlation measurements.

By measuring the amplitude and phase of the wave scattered back from the target of interest at a number of points in a plane normal to the direction to the target (direction of propagation), information about the transverse distribution of the scatterers constituting the target is obtained.

The results of a simulated calculation are shown in Figure 5.14. Here the object is illuminated 3000 m away by one single 10 GHz radio wave (wavelength is 3 cm). We measure the transverse spatial correlation of the received field. Note the marked effect of object shape and the dramatic effect of size even at a distance as great as 3000 m and for radio frequencies as low as 10 GHz.

In an operational system, the target illuminator will shine a small fractiaon of its power on the array antenna. In practice it may be more practical to shift the frequency of the wave illuminating the array slightly relative to the frequency of the wave illuminating the target in order to obtain a suitable intermediate

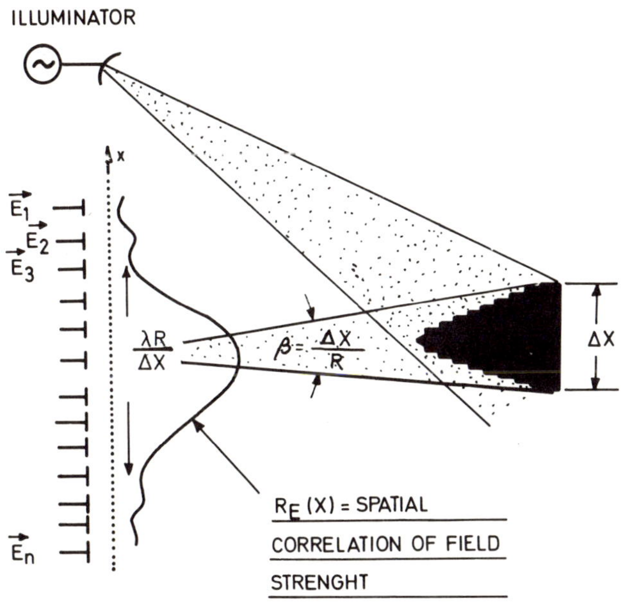

Figure 5.13 Wavefront analysis by array matrix. Spatial correlation of field $R_E(X)$ = Fourier transform of angular distribution of scattered power $P(\theta)$.

frequency for efficient amplification and subsequent quantization and storage in the receiving system. By this system the amplitude and relative phase are measured in each antenna element. An on-line computer processing system (parallel processing) reads the information about amplitude and phase stored at each antenna element and performs the appropriate computations. The reader who is familiar with the principle of holography will immediately recognize this as basic microwave holography.

We shall now present two examples illustrating the potential of the technique for optimum design of practical systems.

5.4 APPLICATION OF REMOTE PROBING TECHNIQUES TO THE STUDY OF THE SEA SURFACE STRUCTURE: AN EXAMPLE

As an illustrative example of the optimum detection/identification technique discussed above, let us consider the structure of the

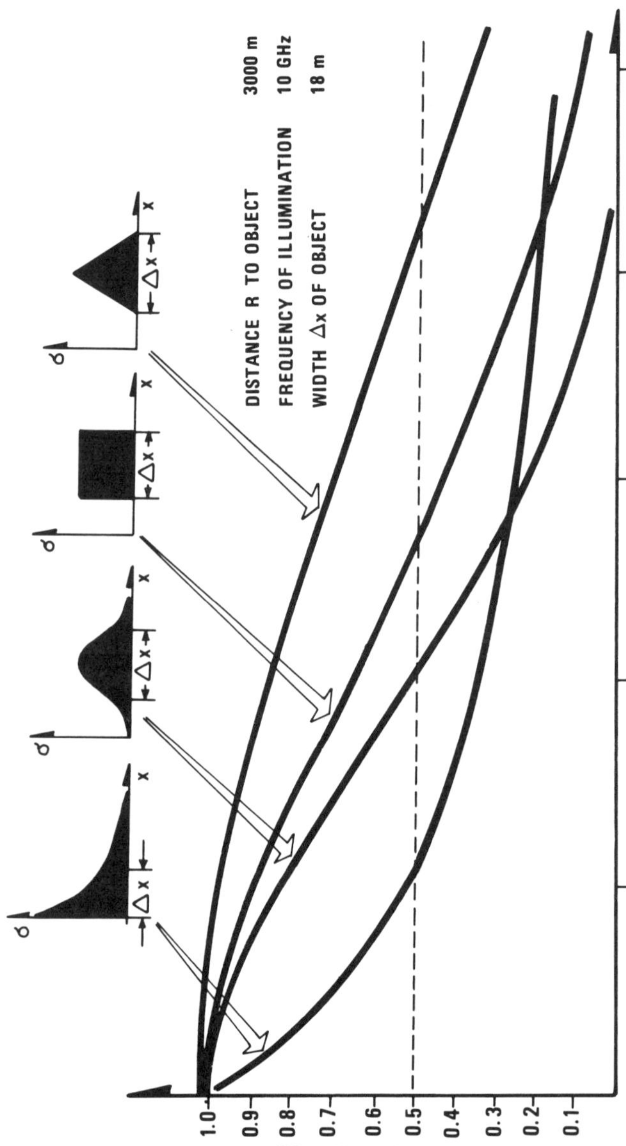

Figure 5.14 Theoretical results illustrating the identification potential of a single frequency illuminator and a multielement-receiving antenna system.

sea surface. Our aim is to determine the directional (two-dimensional) wave height spectrum using optimum remote probing methods. In short, we have learned that:

- The correlation between waves of different frequency is determined by the Fourier transform of the distribution in depth (range) of the scattering elements constituting the scattering object [*i.e.*, the autocorrelation in the frequency domain $R(\Delta F)$ = Fourier transform of the delay function $P(\tau)$].
- The correlation between waves of equal frequency but measured at different points in space is determined by the Fourier transform of the angle of arrival spectrum, *i.e.*, determined by the Fourier transform of the distribution of the scattering elements along directions that are orthogonal to the direction of wave propagation.

In the current example, where our interst is focused on the sea surface structure, we shall make use of waves having different frequencies.

Before going into detail regarding the measuring techniques, let us briefly present the main characteristics of sea waves. In Chapter 3 on scattering from rough surfaces, we established some basic relationships regarding the statistical properties of such surfaces. We have introduced the wavenumber spectrum of the irregularities and have discussed its Fourier transform, the spatial autocorrelation of the surface structure.

For gravity waves in the ocean (waves whose motion is controlled by gravitational forces), the dispersion relationship for deep water is a very simple one:

$$\omega = \sqrt{gk} \tag{5.18}$$

where the wavenumber $\vec{k}$ is two-dimensional, one where the modulus $k = \sqrt{k_x^2 + k_y^2}$. Since $\omega = 2\pi/T$ where T is the period of the waterwave, and since $k = 2\pi/L$ where L is the wavelength of the wave, the relationship between wavelength and period becomes

$$L = \frac{g\,T^2}{2} \tag{5.19}$$

Writing now the waveheight directional spectral density as $S(\vec{k})$ so that the mean square height of the ocean surface is obtained by double integration of $S(\vec{k})$ from $-\infty$ to $+\infty$, we obtain the temporal waveheight spectrum $S(\omega)$ familiar to oceanographers [27-29].

Figure 5.15 shows some measured and model waveheight temporal spectra. Notice the sharp cut-offs at low frequency. The cut-off corresponds to the phase velocity of the waterwave being equal to that of the exciting wind. With increasing wind, the cut-off therefore moves to lower frequencies. A complete directional spectrum could consist of curves similar to these for each azimuth direction.

Imagine now a beam of electromagnetic radiation being pointed at the ocean at a grazing angle γ. The sea surface can be considered as a superposition of diffraction gratings, corresponding to the different components in the directional wavespectrum. If the incident electromagnetic wavevector is $\vec{K}$, the only water wavevector $\vec{k}_o$ that will give constructive interference in a scattered direction $\vec{K}'$ is given by the *Bragg condition*

$$\vec{K}' - \vec{K} = \pm N\vec{k}_o \tag{5.20}$$

where N is a positive integer.

Let us consider the particular case of backscattering, with $\vec{K}' = -\vec{K}$. From the condition above, the only ocean waves contributing to the backscattered signal are those propagating directly away from or towards the antenna, and whose wavelengths satisfy

$$L = \frac{N\lambda}{2\cos\gamma} \tag{5.21}$$

where λ is the electromagnetic wavelength. Moreover, to the first order and for N = 1, the backscattered power is directly proportional to the waveheight directional spectral density

$$P \cong S(k_{ox}, k_{oy}) \tag{5.22}$$

Figure 5.16 shows the Doppler spectrum of sea backscatter via the ionosphere, at 26 MHz. The sea area illuminated was about 2700 km away. From Equation 5.21, ocean waves of wavelength N times 6 m propagating towards or away from the receiver will contribute to the backscattered signal. The Doppler shifts of the first-order backscatter in Figure 5.16 (± 0.52 Hz) give a velocity of ± 3 m/s, which is indeed the phase velocity (*i.e.*, ω/K) of 6-m water waves. Notice how negative Doppler shifts predominate: the wind in the illuminated area was moving predominantly away from the antenna. Notice also the higher-order contributions, clearly noticeable for N = 2, 3 and 4, with the corresponding

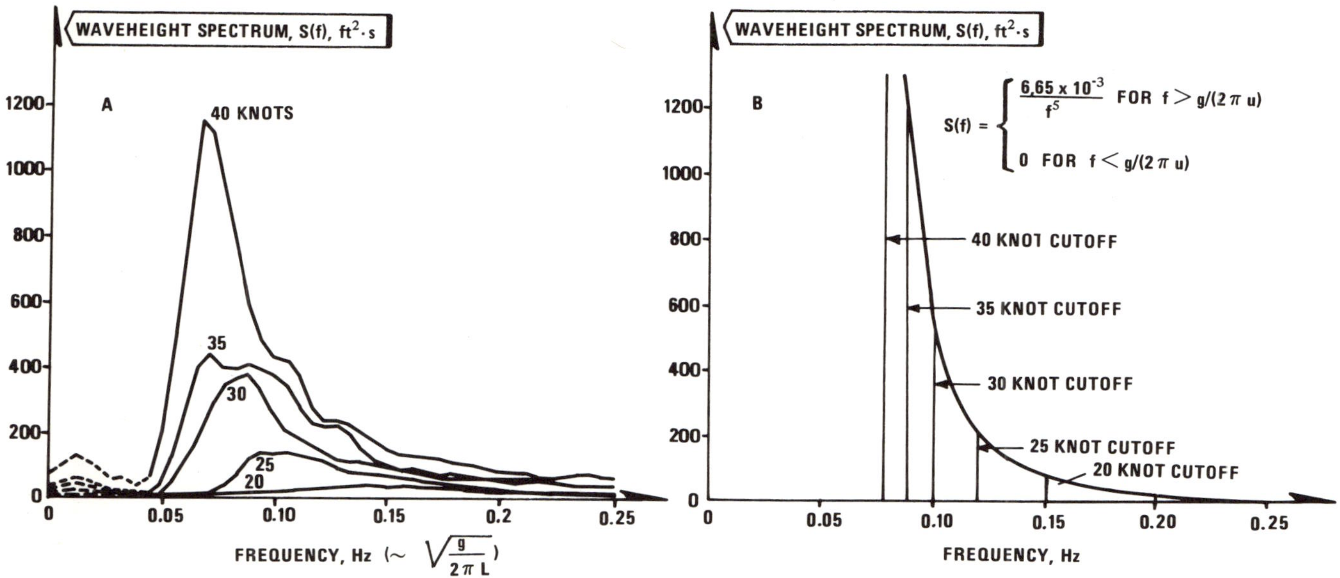

Figure 5.15 Measured and model waveheight temporal spectra. A. Measured (after Moskovitz, 1964); seas are not fully developed at winds about 25 knots. B. Phillips model for fully developed seas.

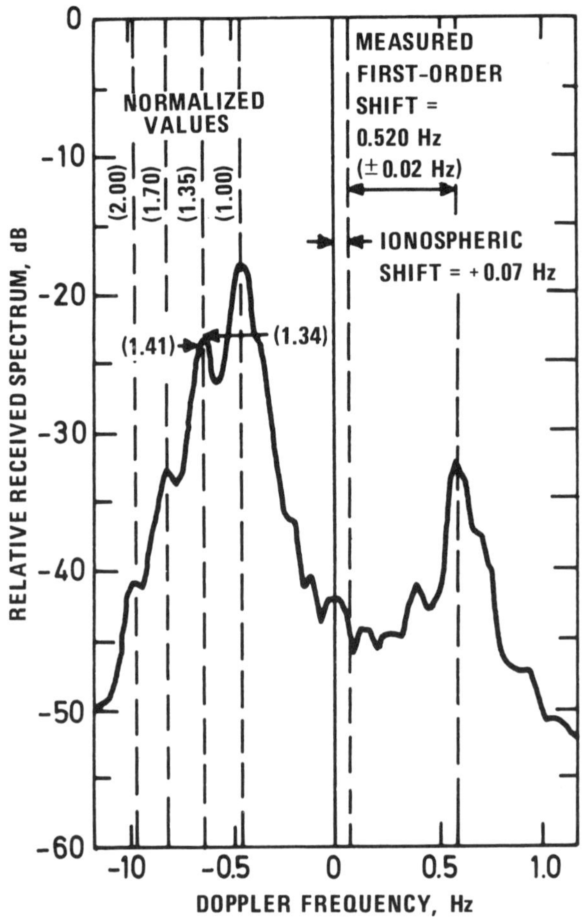

Figure 5.16 Ionospherically propagated sea backscatter spectrum at 25.75 MHz. After Barnum, 1971, see [27].

Doppler shifts increasing as $\sqrt{N}$, in agreement with the gravity wave dispersion relationship, Equation 5.18.

Thus, in principle, the directional wave spectrum can be measured from the return Doppler spectrum for different frequencies. In practice, however, this is nearly impossible because of the antenna dimensions required: a 5° beam surveying 50-m gravity waves would necessitate a 1-km antenna.

Microwaves of frequency 3-30 GHz give narrow beams with antennas of modest dimensions (~ 1 m). They couple to the capillary waves (ripples) of wavelength 1-2 cm, and not directly to the large gravity waves. However, the gravity waves modulate the capillary waves return (Figure 5.17). Referring to Figure 5.17: the sea surface was illuminated with a set of waves having different

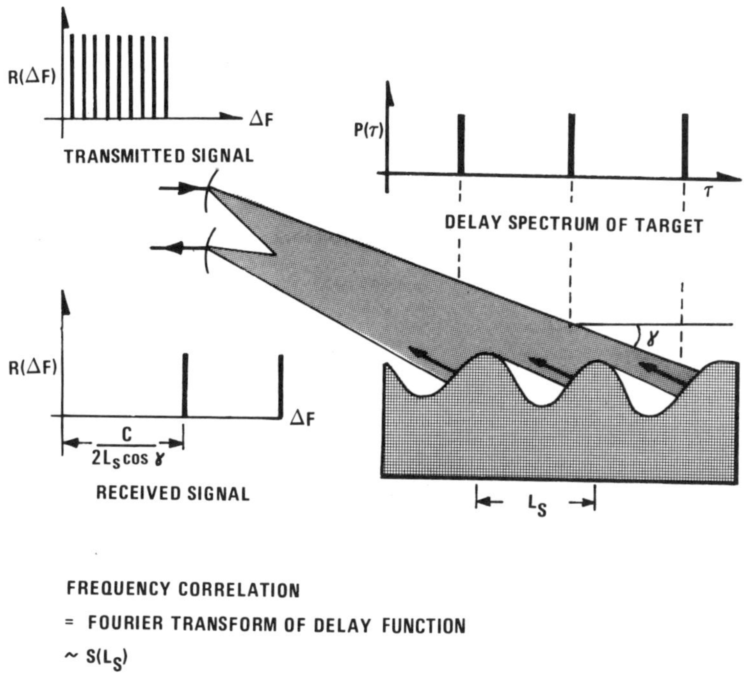

Figure 5.17 Frequency correlation of return signals after illuminating a sea surface with a set of frequencies simultaneously.

frequencies, but which were mutually correlated. An example of such waves is an amplitude-modulated carrier with the carrier frequency removed. The two resulting sidebands are then correlated in amplitude and phase. If the sea surface is sinusoidal, the scattering elements are distributed periodically and the delay function $P(\tau)$ becomes a set of equally spaced delta functions in the limiting case. Thus the correlation properties of the received signal is the Fourier transform of the set of delta pulses, which is another set of delta pulses. As we have seen from Chapter 3, which discussed the basic properties of scattered waves, if L_S is the wavelength of the sea surface waves, the distance ΔF between successive maxima of our correlation function is given by

$$\Delta F = \frac{w\,c}{2L_S \cos\gamma}$$

where c is the velocity of light and w is a constant that depends on the way in which the scattering cross section $\sigma(z)$ decreases with increasing range z. If, as in the current example, we assume that the contribution is the same from each wave facet on the sea surface, then w is unity. Thus, if we are interested in waves of wavelength L_S only, then we should transmit two waves separated by a frequency ΔF given by

$$\Delta F = \frac{c}{L_S \cdot 2\cos\gamma}$$

Note that it is only the difference frequency which matters. In order to obtain the directional (angular) resolution of the sea wave spectrum required, it may be appropriate to use microwaves in the 10 GHz region so as to obtain a directivity of about 4° with small (aircraft-mounted) antennas. The basic principle for this "frequency difference matching" to a reflecting surface is illustrated in Figure 5.18, which depicts a scattering body consisting of two scattering elements only, a distance $\Delta\ell$ apart. The phase angle between the waves reflected from surface 1 and surface 2, respectively, is then

$$\phi = 2\pi \frac{2\Delta\ell}{\lambda_1} = 2\pi \frac{2\Delta\ell F_1}{c}$$

where λ_1 and F_1 are wavelength and frequency, respectively, of the illuminating radio wave.

For constructive interferences, we must change the frequency F so that the phase ϕ is changed through 2π. That is

$$2\pi + \phi = 2\pi \frac{2\Delta\ell F_2}{c}$$

and

$$(F_2 - F_1) = \Delta F = \frac{c}{2\Delta\ell}$$

It then only remains to present the general situation, which is illustrated in Figure 5.19. Knowing the range of pertinent sea

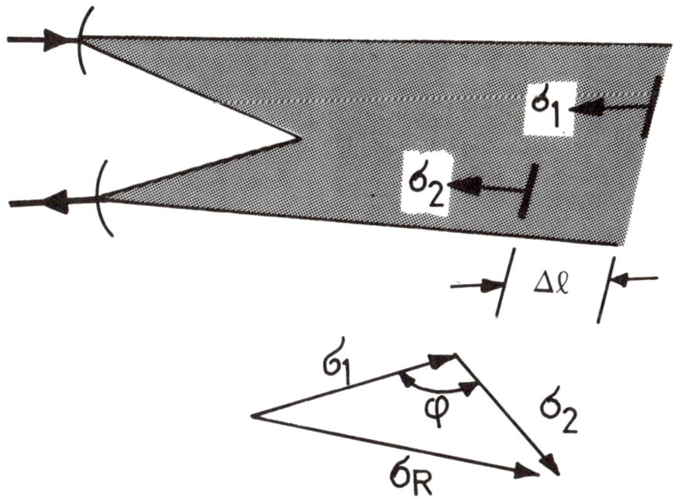

Figure 5.18 Scattering from sea surface, simplified principle.

surface wavelengths, we know the maximum and minimum frequency spacing of our two radio waves, the correlation of which is to be measured. Having measured the correlation (covariance) over the frequency range of interest, the amplitude wavelength spectrum of the sea surface is deduced.

Chapter 6 will present some detailed information on recent experiments. For details the reader is referred to [28].

5.5 APPLICATION OF REMOTE PROBING TECHNIQUES TO THE STUDY OF WATER POLLUTION: AN EXAMPLE

We have seen that in order to obtain the highest sensitivity and selectivity of a spectrometer system, the illumination has to be structured in a very particular manner. This introduced the concept of "matched illumination." Similarly, in order to optimize the conditions at the receiving end, we shall apply a "filtering method" to the received signal. In order to illustrate the effect of a "matched illumination" together with "matched reception," we shall use a fluorescent medium as an example.

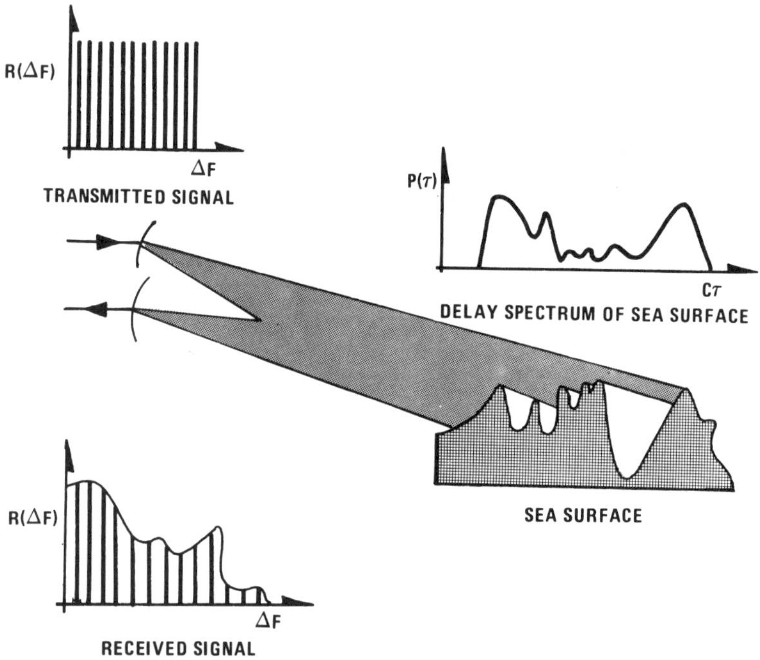

$$S(L_S) = FT(R(\Delta F))$$

Figure 5.19 Scattering from sea surface, general principle. Microwaves give directional wavespectrum.

In addition to serving as an illustration of the concept of double matching, the study of water pollution by laser fluorescence methods is of interest on its own merits. It is well known that water pollution influences the ecology of the algae in the sea and in inland waters. Specifically, from knowledge about the distribution of the various algae populations, information about the purity of the water can be deduced. It may, for example, prove difficult to measure directly the concentration of a particular pollutant in a lake by remote probing methods, but it may, as will be seen from the example to follow, be possible to measure the concentration of a particular algae (*e.g.*, the *Oscillatoria bornetii*) that appears to be influenced by chemicals in the water (phosphates, nitrates, iron).

Figure 5.20 illustrates some of the fundamental properties of fluorescent biological substances. Four different algae are considered, each characterized by an excitation spectrum (left-hand diagram) and by an emission spectrum (right-hand diagram). The

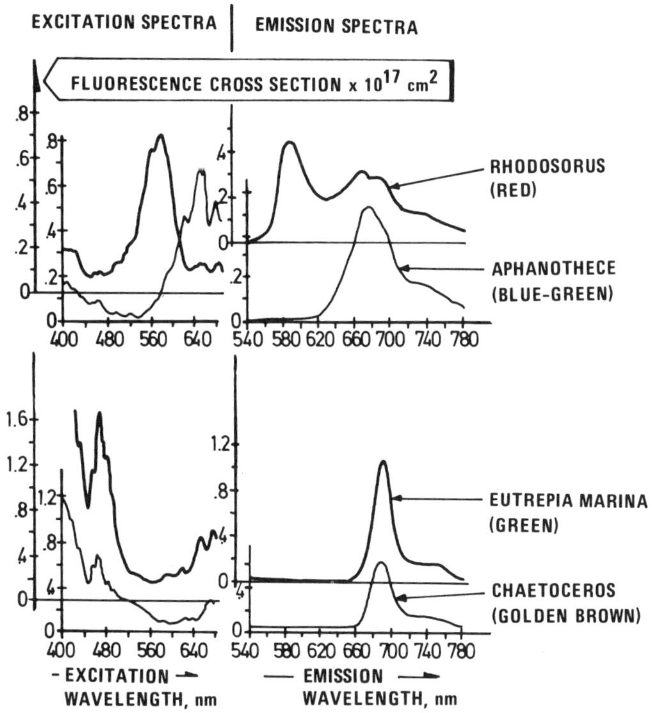

Figure 5.20 Example of excitation and emission spectra (after P. M. Mumola).

excitation spectrum gives information about the efficiency of the fluorescence process as a function of excitation wavelength, whereas the emission spectrum is the wavelength distribution of the emitted light.

Making use now of the arguments presented in Section 5.3, we see that in order to maximize the stimulation of the particular algae species of interest, we illuminate the algae with light having a very particular wavelength distribution. Particularly, we switch on a set of monochromatic light sources having different wavelengths and different intensities in a particular sequence so as to achieve a periodic sinewave variation in the fluorescent light intensity requiring a minimum of detection bandwidth (maximum integration time). At the receiving end, a set of light filters is followed by light detectors and narrow band amplifiers. At the output of these amplifiers a receiver processor correlates the spectral distribution of the fluorescent light with the distribution of the particular species of interest. By this process maximum sensitivity and species selectivity are achieved. Figure 5.21 illustrates the particular experimental technique.

Environmental parameters that can greatly affect the fluorescence cross section of algae include long- and short-term light history and nutrient and age effects. This calls for rather extensive and frequent local "ground-truth" measurements in order to utilize the technique fully.

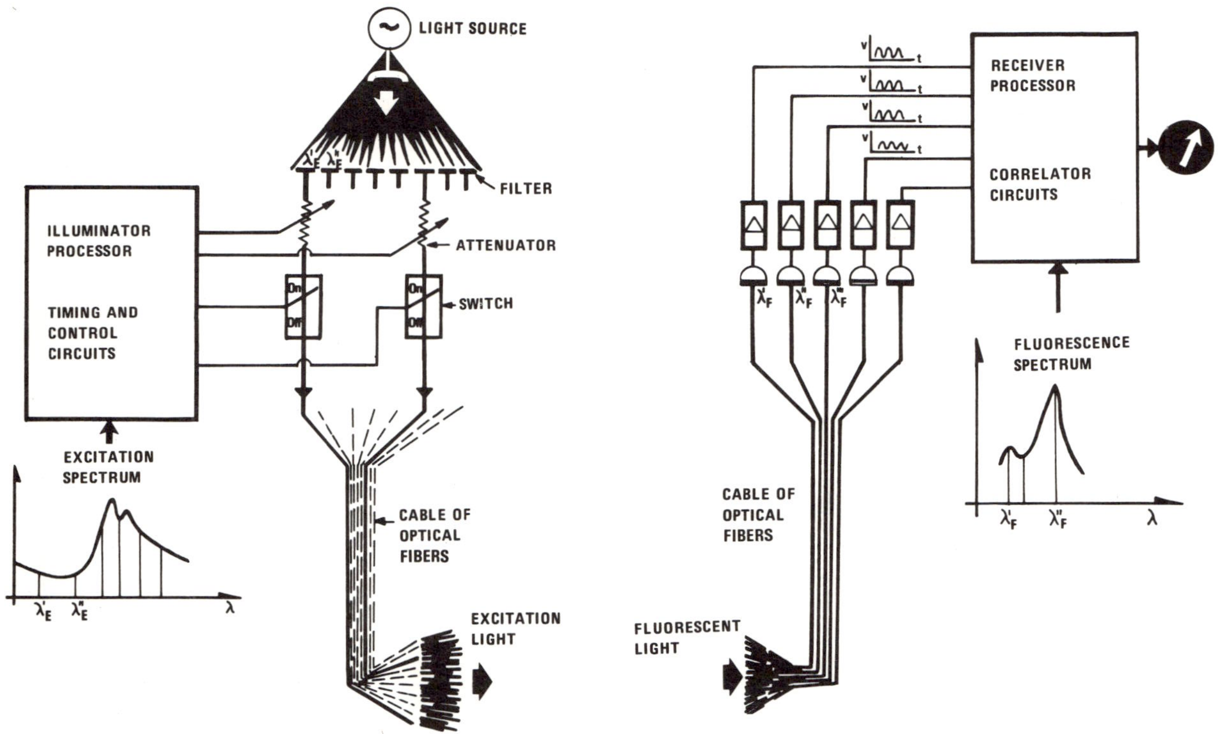

Figure 5.21 An example of selective detection of algae using "matched illumination" and "matched reception."

CHAPTER 6

PRESENT STATE-OF-THE-ART: A DESCRIPTION OF A SET OF ENVIRONMENTAL SURVEILLANCE METHODS

In Chapter 5 on methodology, we discussed general asepcts of remote sensing. On the basis of a set of methods developed in the field of radio science, we have presented a methodology for remote sensing of environmental factors. To complete the picture, a representative set of available environmental surveillance methods must be presented. Obviously, to give a comprehensive account of the techniques available today would require a great deal of space. The objective of this chapter is to give a condensed review of a selected set of important techniques. This does not necessarily imply that a technique which is important from the user's point of view, in that it solves important problems, will be given the greatest emphasis. The selection is made on the basis of the novelty of the technique and on the degree to which it is exciting to the author. Clearly then, the selection will be a subjective one; some readers will no doubt search in vain for techniques they consider important. The author hopes, however, that only a few of the interesting remote probing methods have been omitted and that the reader, like the author, will be fascinated by most of those mentioned.

This author believes that the progress in the field of environmental surveillance by radio science methods is remarkable and most stimulating, probably because there is an urgent demand for such methods and at the same time the means to satisfy the demand have been acquired. The tools at our disposal are, to a large extent, inherited from the radio scientist who addressed himself to problems in the troposphere (radio meteorology) and beyond into space.

Finally, it should be noted that the brief survey in this chapter is not the result of a comprehensive literature study. It is based largely on personal interviews with individual scientists who are experts in this field. However, it is hoped that the chapter will stimulate the reader and that it will be of aid to environmental researchers who are seeking powerful methods for environmental surveillance.

6.1 SEA SURVEILLANCE

Remote probing techniques aimed at measuring the conditions at or in the immediate vicinity of the sea surface will be considered.

6.1.1 Wind Close to Sea Surface

1. *Multifrequency Radiometry*
   (Dr. T. T. Wilheit, NASA Goddard Space Flight Center)

A nearly all-weather capability for measuring wind speed at the air/sea boundary can be obtained using a set of multifrequency radiometers at microwave frequencies. There are several factors affecting the radiance as viewed with microwave radiometers: the sea surface temperature, the roughening of the sea surface by the wind, and the composition and temperature of the intervening atmosphere. Each of these factors has its own characteristic spectrum and polarization properties.

This principle is to be exploited in the scanning multichannel microwave radiometer (SMMR) to be flown on the SeaSat-A and Nimbus-G satellites in 1978. These instruments will measure the microwave radiance at 6.6, 10.7, 18, 21 and 37 GHz at a nadir angle of 48° and in two linear polarizations. Sea surface wind accuracies of better than 2 m/s at standard height (19.5 m) are expected with a resolution of about 100 x 100 km. Aircraft-borne radiometric measurements of wind speed have been reported by Wilheit and Fowler and Webster *et al.* [30,31].

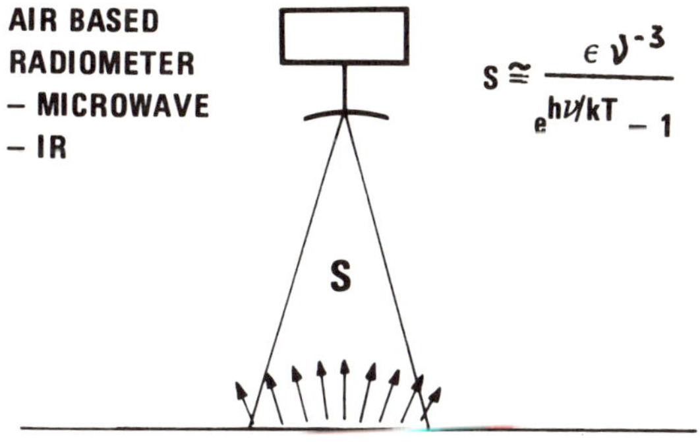

Figure 6.1 Principle of the radiometer for sea surface investigations. Emission coefficient ϵ is determined by surface structure and surface chemistry. Relative dependence varies with frequency. T is temperature and ν radiation frequency.

2. *Scatterometer*

In addition to the radiometer techniques, an active scatterometer is proposed for the measurement of surface wind speed. Whereas the radiometer system is most sensitive to wind speed in excess of 5 ms^{-1} (it responds to foam cover), the scatterometer is useful for wind speeds in the range 3-15 ms^{-1}. A combined scatterometer and radiometer will be flown in NASA SeaSat-A in 1978 [32].

The scatterometer operates at 14.6 GHz and illuminates the sea surface. The interference patterns received are analyzed in much the same way as in previous measurements of ionospheric irregularities. From this, wind speed can be deduced (range 3-15 m/s). The scatterometer illuminates the sea surface with four lobes, two lobes on either side of the spacecraft. By means of these sets of lobes information about the degree to which the sea surface motion is "frozen" can be measured.

6.1.2 Sea Surface Temperature (Dr. T. T. Wilheit, NASA Goddard SPFC)

Microwave radiometers at five frequencies (6.6, 10.7, 18, 21 and 37 GHz) and two polarizations give a set of information on the basis of which sea surface temperature is obtained. The accuracy in a satellite application is expected to be 1.5°K. The system offers all-weather capability. The sea surface resolution cell will be determined by the lowest frequency (6.6 GHz) and will be approximately 150 x 150 km. The five-frequency radiometer is based on a direct mixing superheterodyne receiver (*i.e.*, no RF preamplifier) with a noise factor of 5 dB. Long period averaging of received signals is essential. The sea surface temperature is essentially that at the surface, down to a millimeter or so. The reflection and transmission properties of the atmosphere must be taken into consideration (the emissivity of the sea at these frequencies is approximately 0.5, hence the reflectivity is also 0.5) [31].

Sea surface temperature is commonly measured by high resolution IR radiometry in the atmospheric window regions. Figure 6.2 illustrates its high variability, spatially and in time [33].

6.1.3 Sea Surface Salinity (Dr. B. Kendall, NASA Langley Research Center)

NASA Langley (Bruce Kendall and H-J. C. Blume) measures sea surface salinity and temperature by the use of two airborne microwave radiometers (S- and L-band, respectively). The brightness temperature (power received at the radiometer) is a function of molecular temperature and salinity. In the S-band (2.65 GHz), temperature is the dominating parameter, whereas salinity dominates in the L-band (1.43 GHz). Using both frequencies and compensating for the influence of cosmic radiation, intervening atmosphere, sea surface roughness, and antenna beamwidth, accurate information about both temperature and salinity is obtained. The spatial resolution on the sea surface is approximately 3/10 of the height of the observation platform (the aircraft). Demonstrated absolute accuracy (by comparison with sea truth measurements) is 1/1000 for salinity and 1°K for temperature [34,35].

6.1.4 Atmospheric Surface Pressure Over Sea (Dr. G. E. Peckham, Heriot-Watt University)

A method making use of the atmospheric transmission within the 5-mm oxygen absorption band is suggested by G. E. Peckham

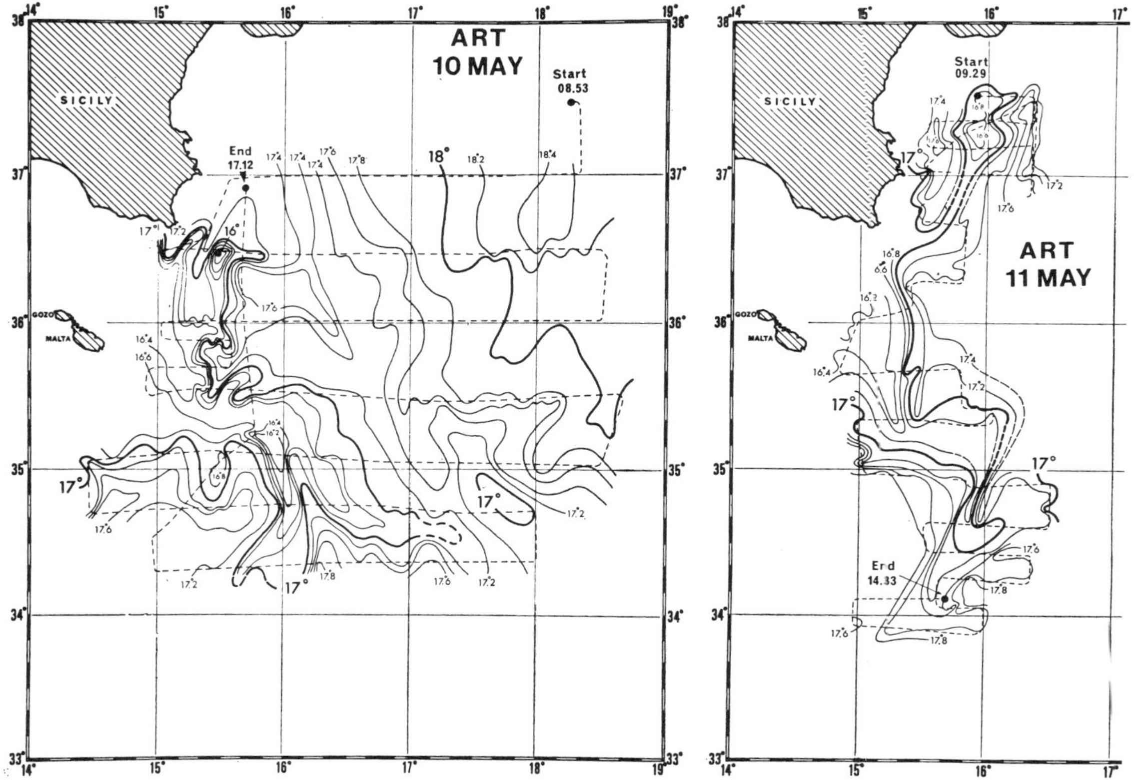

Figure 6.2 Corrected sea surface isotherms from ART measurements, May 10 and 11, 1972. After Briscoe, Johannessen and Vincenzi [33].

and D. A. Flower, Heriot-Watt University, Scotland (information from NASA). Intensities of signals at several frequencies in the 50 GHz region as reflected from the sea surface are used to allow for the variation in surface reflectivity and cloud and water vapor transmissivity. The method suggested could, when referred to accurate *in situ* measurements of pressure at certain points on the surface, give global coverage. Supplemented with information about atmospheric temperature as obtained by passive IR and microwave radiometers, the entire atmospheric pressure field can be deduced using the equation of hydrostatic balance. Based on a rather crude instrumentation (source power 3 W, antenna 30-cm diameter, receiver bandwidth 100 kHz, noise factor 10 dB) there appears to be no fundamental reason why ± 3 mb should not be achieved [36].

6.1.5 Precipitation Over Sea (Dr. T. T. Wilheit, NASA Goddard SPFC)

Microwave radiometry viewing the sea surface from above (Nimbus 5) at 19.35 GHz gives information about rain rate with accuracy of about a factor of 2. The principle on which the method relies is emission from precipitation using the cold sea surface as reflector.

6.1.6 Surface Structure of the Sea

Present techniques for measuring sea state are centered on the use of HF radar for measuring sea state over an area of approximately 2 x 10 km^2 from a single station. It is anticipated that systems will be operational in the near future. Satellite techniques (see Section 6.3.2 on wind speed) based on combined microwave radiometer/scatterometer observations are under investigation, but considerable research is still needed, and a 10-to-15-year effort may be anticipated before this evolves into an operational system. About two years of experimental data gathering are required to complete the development stage of the HF system and to enable a decision on whether to implement it operationally. The two techniques are complementary in that the HF system has been shown to be cost-effective, even on a short-term basis; hence it could fill in the gap before the satellite system becomes operational [37,38].

1. *Microwave Altimeter*
(Dr. A. Loomis, JPL)

NASA SeaSat-A will be performing studies of the general ocean current and dynamics. From these studies (dynamic topography) ridges as high as 1.5 m above mean sea level are to be expected (D. W. Van Arx, Woods Hole).

An altimeter designed at JPL has the following characteristics:

- Frequency is 13.9 GHz
- Pulse compression techniques give a range resolution of 10 cm
- Beamwidth enables spatial resolution of 1-km diameter on the sea surface. Taking the motion of the satellite into account, the resolution cell is 1 km x 7 km.

This system gives good waveheight statistics and absolute height calibration.

2. *Synthetic Aperture Radar (SAR) Imager*
(Dr. A. Loomis, JPL)

The SeaSat-A synthetic aperture radar imager has been mounted in a 990 Convair aircraft and has the following characteristics:

- Frequency is 1.275 GHz
- The system is focused, *i.e.,* it corrects for wavefront curvature
- Integrates up to one second (300-1000 pulses)
- Looks 20° from the vertical
- Resolution is 25 m in both directions
- Phase resolution is 3°
- Range to object (sea surface) to be imaged is up to 30,000 ft.

A laser is used with this instrument to measure slope distribution of ocean waves [32].

3. *Land-Based HF Radars*
(Dr. D. E. Barrick, WPL, NOAA)

The use of shore-based radars that propagate via the ionosphere appears to be an ideal solution to the problem of wide-area routine sea state monitoring. For example, a single backscatter radar located in Kentucky could observe sea scatter over the entire Atlantic coast out to a range of about 1200 nautical miles. The dominant (first-order Bragg scattering) contribution to the echo from the sea surface at MF and HF arises from the ocean waves whose wavelength is exactly half that of the radar wavelength. Since these deepwater ocean waves have a phase velocity that bears a direct relationship to the wavelength L of the ocean wave ($v = \sqrt{gL/2\pi}$, where g is the acceleration of gravity), there is a particular Doppler shift Δf associated with the particular ocean wavelength L. This Doppler shift is given by

$$\Delta f = \pm\sqrt{\frac{g}{\pi\lambda}}$$

where λ is the HF wavelength since $L = \lambda/2$.

Finally it has been shown [37,38] that the strength of the first order echo (associated with a wavelength $L = \lambda/2$ and a Doppler shift Δf) is proportional to the height of the ocean wave. The following accuracies are expected: wave direction ± 15°, RMS height error 10%, wavelength 15%.

In order to obtain the whole spectrum of ocean waves, the HF frequency will be swept over an appropriate frequency range. The effect of the ionosphere is noticeable, all the frequency range to be used is obtained on the basis of ionograms. The influence of the ionosphere is under investigation. An example of a surface wave spectrum thus obtained is shown in Figure 6.3 [29,39].

4. *Two-Frequency Microwave Radar*

The principles of the multifrequency technique were explained in Section 5.5. Dr. W. J. Plant, of the Naval Research Laboratory, Washington, D.C., has recently published experimental results using this technique. He used a coherent, dual-frequency CW X-band radar. The frequency difference Δf between sidebands could be adjusted between 7 and 150 MHz. The power spectrum of $E_1 E_2{}^*$ (product of the two detected signals) shows a low-frequency peak corresponding to the appropriate ocean wave component, plus a broad background (~ 100 Hz width) corresponding to the convolution of the single-frequency returns. The height of the low-frequency peak is a measure of the amplitude of the corresponding water wave. Figure 6.3 shows the wavespectra inferred. They correspond roughly to those determined by a wave gauge.

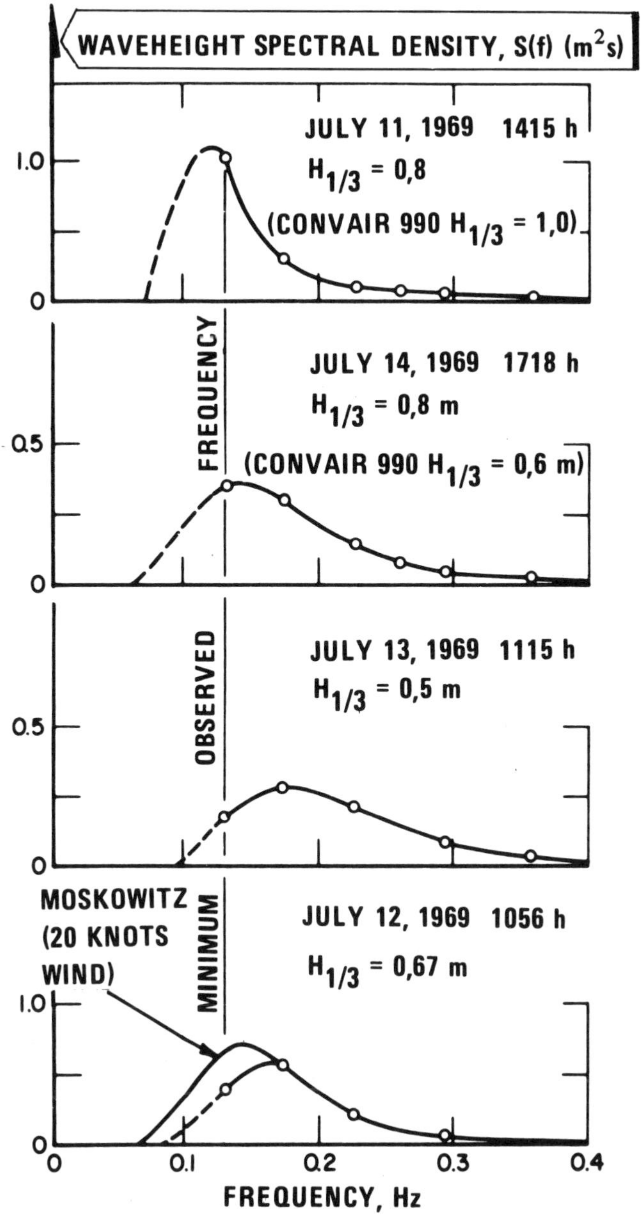

Figure 6.3 Temporal waveheight spectra deduced from MF/HF ground wave backscatter observations. After Crombie, 1971 [39].

Before the multifrequency principle can be applied to routine ocean observations, a better quantitative understanding must be gained about what influences the magnitude of the modulation signal. The primary backscatter comes from the capillary waves. Their interaction with the long gravity waves varies strongly. Thus, in Figure 6.4b the open circles indicate measurements after a change in wind direction. This change did not influence the gravity waves, so the wave spectrum component studied did not change. The modulated signal, however, increased by more than a factor of 2 [27,28].

6.1.7 Sea Surface Currents (Dr. D. E. Barrick, WPL, NOAA)

Surface currents cannot be determined today by using the ionosphere as the reflector. Direct waves (ground waves), however, in the HF region, may be used (20 MHz). A small transportable system for rapid evaluation of the signals received out to a range of some 70 km has been developed. The spatial resolution is 3 km. This is obtained by range-gating (20 μs pulse width) and by measuring the phase and amplitude at four array elements of the receiving antenna, thus obtaining direction. It takes two different stations placed along the coast of interest to obtain the magnitude and direction of the current. The ocean coverage with a baseline of 40 km is 70 x 70 km (see Figure 6.5).

The system, which is referred to as CODAR (Coastal Ocean Dynamics Application Radar), has a peak transmitted power of 2.5 MW (average power 50 W) and the beamwidth is 180° with a three-element Yagi antenna. The system is controlled by a minicomputer (PDP 1134). To obtain good ground plane conditions for surface wave propagation, the antenna should be mounted close to the beach. This method enables measurement of the current at the sea surface to a depth approximately equal to $\lambda/8\pi$. With a frequency of 20 MHz this corresponds to 60 cm. The Doppler resolution capability of the system is 1/100 Hz, giving a surface current resolution of about 10 cm/s. These amplitude and current measurements open up the possibility of detailed studies of the dynamics of the ocean.

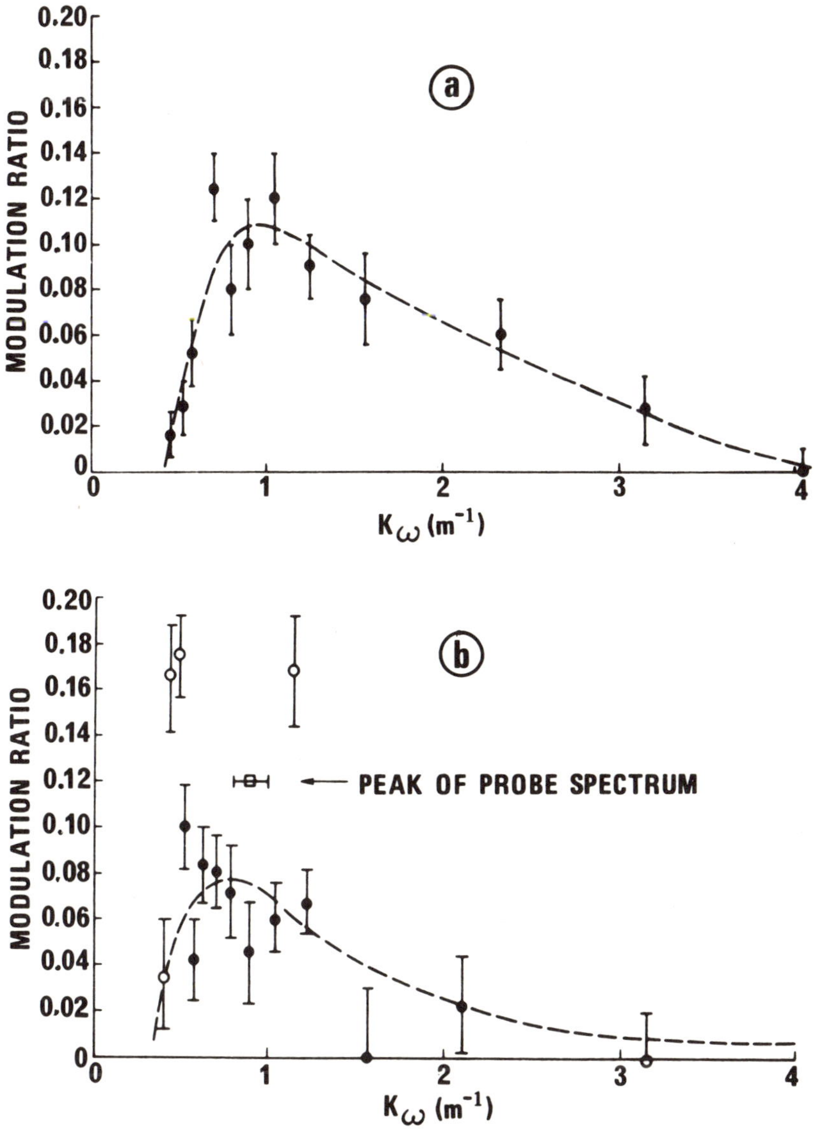

Figure 6.4 Intensity of gravity waves vs water wavenumber. (a) Data of April 10, 1975; (b) Data of June 11, 1975. Dashed lines are empirical. Square indicates wavenumber of peak of probe spectrum; its vertical placement is arbitrary. After W. J. Plant [27].

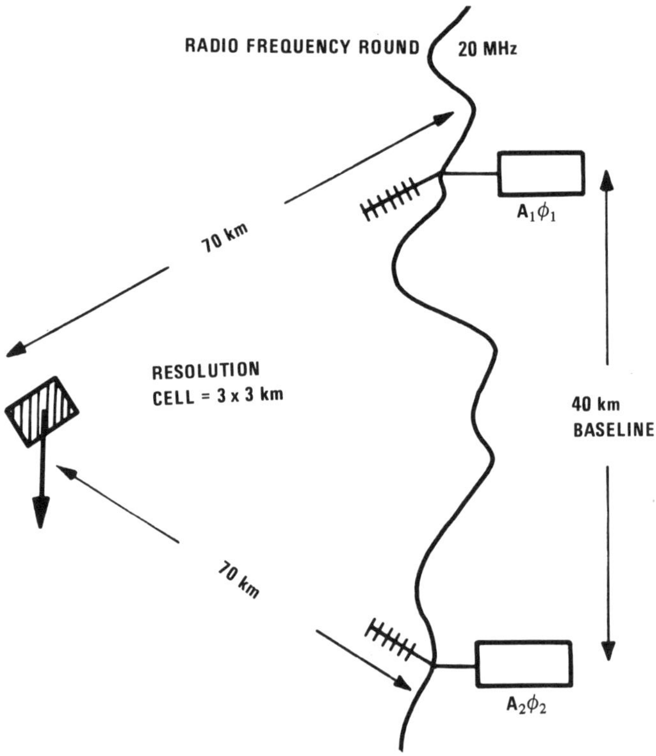

Figure 6.5 Remote probing of sea surface currents. Spatial resolution is 3 x 3 km; current resolution is approx. 10 cm/s.

6.1.8 Water Pollution

Several methods of potential interest will be presented.

1. *The Laser Fluorescence Methods*
   (Dr. M. Bristow, EPA, Las Vegas)

The area under investigation is illuminated with blue or ultraviolet radiation and the reemitted fluorescence is analyzed both in time and spectrally. The method is powerful for many applications:

- Study of plankton, algae and hence the effect of pollution due to domestic use of water, *i.e.,* phosphorus compounds.

- Remote sensing of oil spills: presence, extent and identification and characterization.
- Organic compounds in general in water. The total burden of organic compounds is measured (TOC). The Raman backscatter from water (which is relatively intense) is used as a reference standard with which to compare the fluorescent backscatter from pollutants. The fluorescence spectrum from pollutants in water does not possess a detailed "fingerprint." The spectrum is very broad and featureless.
- Water acidity (the pH value) can have an effect on the fluorescent spectrum. It could be that this is not a direct effect, but rather a secondary one. Acidity affects the organic compounds, which again are analyzed by laser fluorescence spectroscopy.
- For the purpose of tracking the motion of pollutants in a given area (effluent from rivers), fluorescent dyes are added to the water. These can be traced effectively even at very small concentrations with this method. The depth of penetration into the water is determined by the characteristic attenuation length and is generally about one meter. The fluorescence from sea water is much less pronounced than that from freshwater. Examples of spectroscopic measurements making use of the fluorescent mode are shown in Figure 6.6 [40-44].

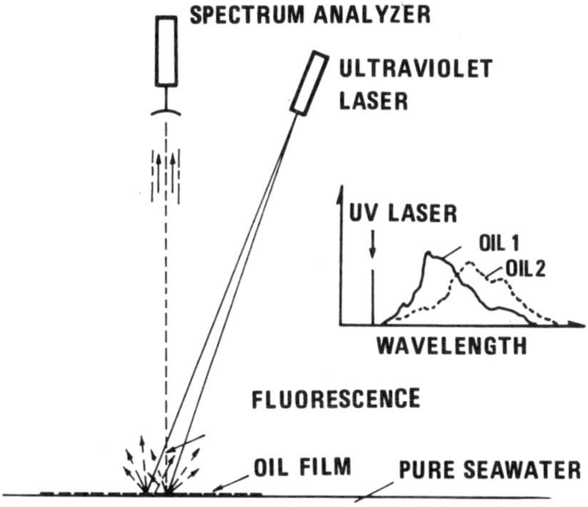

Figure 6.6 Study of surface water pollutants by the use of laser-induced fluorescence.

2. *Detection of Oil Spills by Radar*
(J. Nilsson, Swedish Defense Research Establishment)

It has been established experimentally that relatively small amounts of oil distributed on the ocean surface have a pronounced effect on the ocean capillary waves and hence on the scattering of microwaves from the sea surface. Specifically, modifying the video processor in a conventional airborne weather and ground mapping radar results in the following typical detection performance of the radar.

- With a sufficient quantity of capillary waves (wind speed between 3 and 13 m/s) oil films of thickness of 3 μm can be detected.
- Wind statistics from the North Sea (1970) show that wind speed is less than 3 m/s 7.5% of the total time. For 11.5% of the time wind speed is above 13 m/s. Under such conditions of high wind speed most, if not all, detection methods fail.
- The radar will operate satisfactorily under all cloud conditions. Flying at a height of 3000 ft enables coverage of an area 100 km wide in one flight leg when viewing the sea surface with a depression angle greater than 1°. Until now, maximum detection range for oil has been 20-30 km.

Best contrast between an oil slick and pure sea water is obtained with upwind illumination favoring a side-looking radar configuration. Detection criteria based on polarization and on the time correlation properties and amplitude distribution statistics of the sea echo are being studied by H. Parr and A. Kjelaas, who are working with the author [45-49].

3. *Alternative Methods for Oil Spill Detection*

Ultimately, surveillance of oil spills will be based on multisensor systems because of the many possible interferents and the large variations in environmental conditions. Such an instrumental package may consist of:

- *Laser fluorosensor*
- *High-resolution TV* (an active, range-gated system is being developed for the U.S. Coast Guard by Naval Air Development Center).
- *Infrared linescan.* IR-imagery of oil slicks relies on differences in emissivity or temperature. Temperature differences between oil slick and sea water may be due to absorption of sunlight or vaporization of the lighter oil fractions.
- *Imaging microwave radiometer.* The elimination of capillary waves reduces the apparent emissivity in regions with thin (μm) continuous oil slicks. Thicker portions of the slick will increase the emissivity by acting as an antireflection coating. When the "optical" thickness exceeds a quarter wavelength, the emissivity decreases again, and still increasing thickness will give a periodic modulation of emissivity. A two-frequency radiometer can determine oil thickness up to a

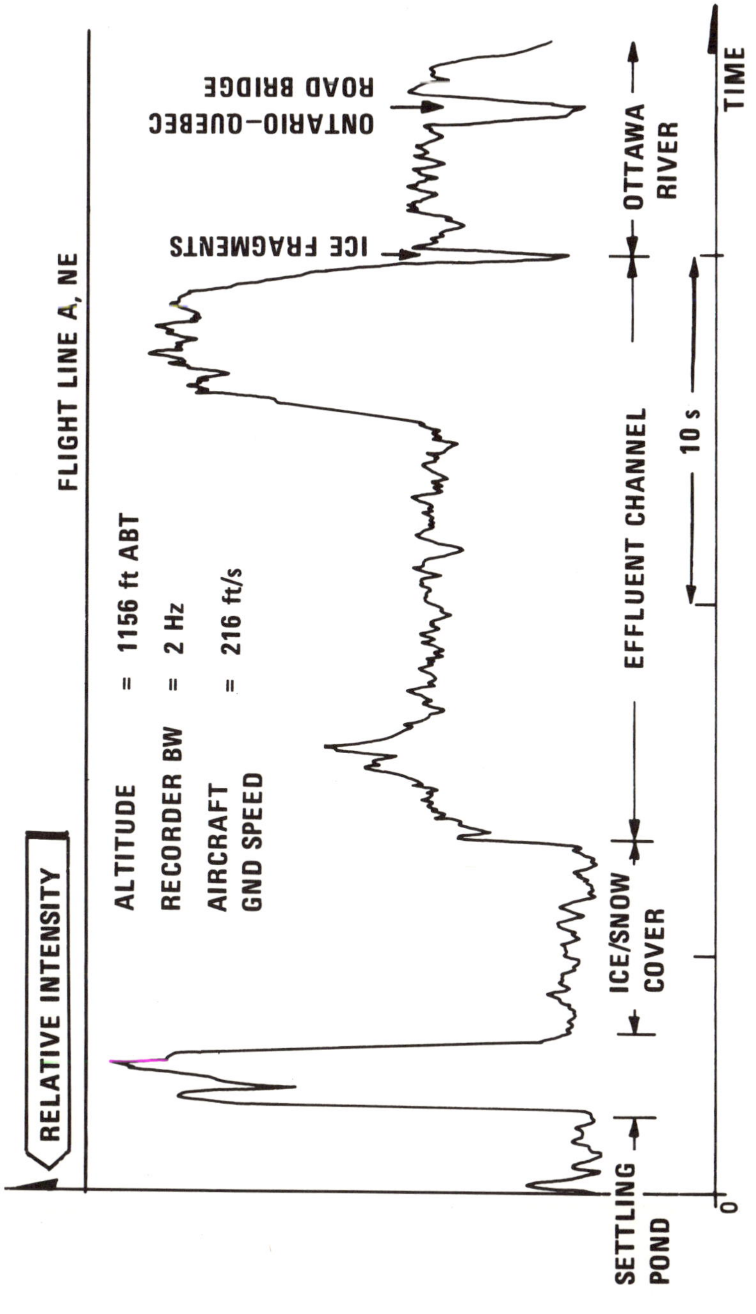

Figure 6.7 Fluorescence profile of pump mill settling pond and effluent channels. After M. Bristow [41].

quarter wavelength with adequate accuracy without previous knowledge about the dielectric constant of the particular oil type.

J. P. Hollinger from the Naval Research Laboratory, Washington, D. C., has led the efforts directed towards practical application of this principle. The NRL practical work is very advanced. Airborne measurements at 19.3 GHz, 31 GHz and 69.8 GHz have been carried out for a series of controlled oil spills [50]. Figure 6.8 shows the apparent temperature recorded when flying over an oil spill. Notice how the thick part of the spill shows up as a strong increase in emissivity, while the thin, outer part of the slick yields a slight reduction because of the missing capillary waves. Recently, the NRL operated an imaging version of the equipment. Coupled with computer processing, this permits rapid contouring of the thickness distribution of the slick to about 0.1-mm accuracy [51].

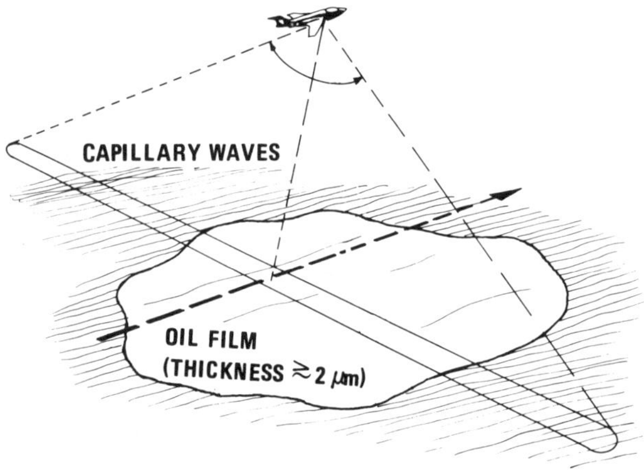

Figure 6.8 Detection of oil spills by side-looking radar (SLAR). The oil film suppresses capillary waves that have a dominating effect on the radar echo.

- *An IR Reflectometer.* This unit is based on characteristic anomalous dispersion in the index of refraction of both water and oil in the 2-4 μm band. Only stationary "across the river" installations are in use [52].

6.2 LAND SURFACE SURVEILLANCE

6.2.1 Land Surface Pollution Detection

Whereas remote surveillance techniques have proven useful in the study of the sea surface and of the atmosphere, the problem of measuring the pollution burden on the land is a much more difficult one. This is, among other things, a result of the multitude of different compounds involved. Under favorable conditions, the effect of pollution on vegetation may be measured by multichannel, wide-band, photographic techniques or scanning (visible to IR wavelength range), like that provided by Landsat. The ultimate endeavor, however, is to measure the geographical distribution of a particular pollutant on the ground and on vegetation. The detection method may be based on the following principle.

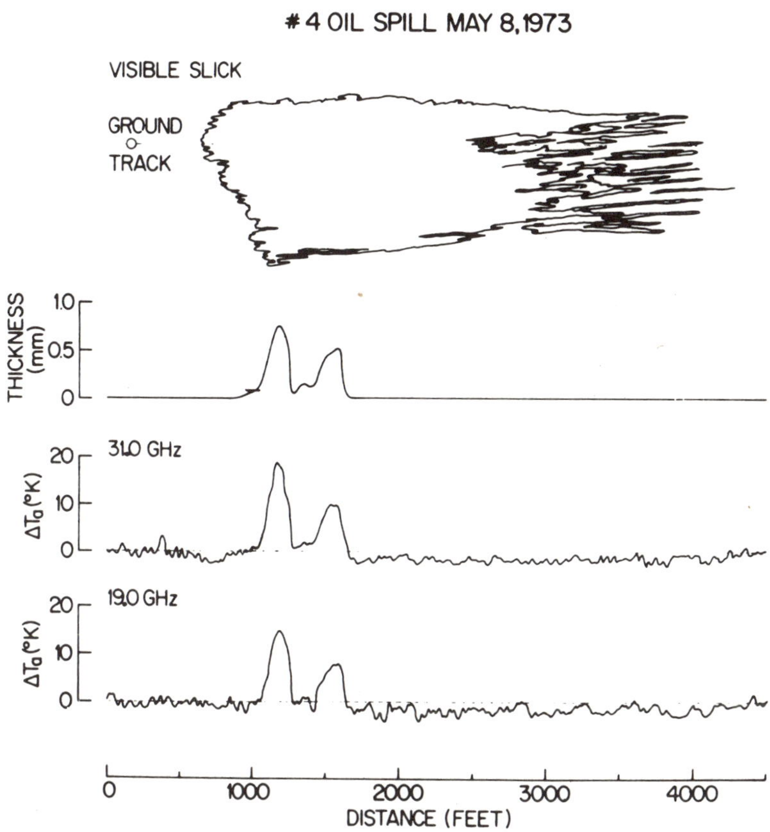

Figure 6.9 Oil slick thickness profiles determined by two-channel radiometry. From J. P. Hollinger [50].

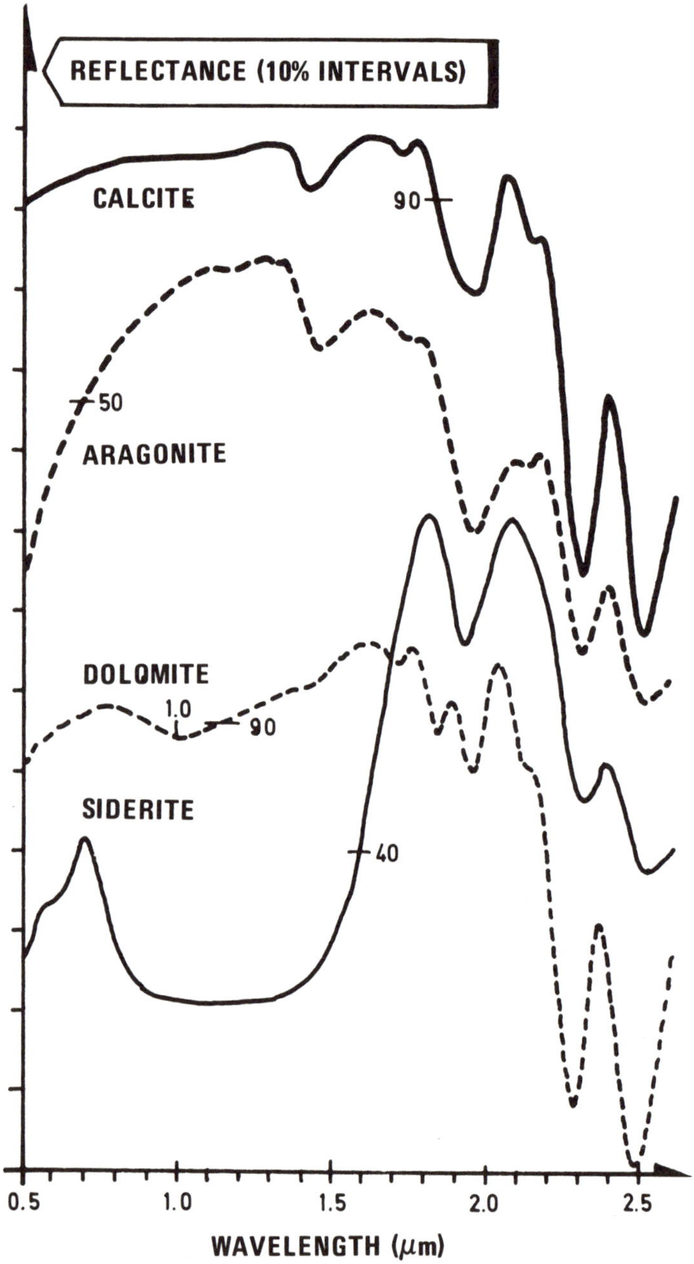

Figure 6.10 Diffuse reflectance spectra of carbonates. Electron (Fe^{2+}) absorptions occur near 1 μm, and strong water and carbonate-ion appear between 1.4 and 2.5 μm.

We know what to look for if we wish to detect a chemical agent whose composition is known so well that its absorption spectrum (or emissivity or reflectivity spectrum) is also known. The problem is to structure the "illumination" so as to make optimum use of the "fingerprints" of the agent of interest, bearing in mind that the agents of no interest (the interferents) have their fingerprints too. If the reflectance spectrum of the agent of interest is sufficiently well known, an optimum illumination can be chosen so as to require a minimum detection bandwidth. Thus, the sensitivity of the system (inversely proportional to detection bandwidth) can be made large for the agent of interest and suppressed for all the presumably unknown interferents, including the effect of the ground irregularities.

The primary signature may be the absorption spectrum of the pollutant itself or some characteristic change in the surface reflection coefficient or the detailed topography (vegetation anomalies) of the surface. Thus the detection method is not linked to any particular sensor system.

6.2.2 Crop Classification

Traditionally, crop classification is a task for multispectral imagery. However, the signatures of features characterizing vegetation species are numerous and some of them well matched to microwave interactions, *i.e.*, textures and height. DeLoor [53] uses a radar backscatter technique. The crop classification is obtained essentailly by measuring the normalized scattering cross section σ_0 as a function of frequency band, angle of incidence (depression angle) and polarization.

6.2.3 Geology

Radar imaging has proven a valuable tool in physical geography and reconnaissance mapping on a regional scale. Rock type may also be determined by studying the emissive (IR and microwave), reflective (including UV) and fluorescence properties.

6.2.4 Hydrology, Snow Cover and Soil Moisture Studies

It is expected that multifrequency microwave radiometric mapping will contribute substantially to a routine mapping of "water inventory." Dr. E. Schanda of the University of Berne, Switzerland, has taken on an extensive study of signatures [54].

6.3 ATMOSPHERIC SURVEILLANCE

The atmosphere (troposphere and stratosphere) is the medium that has been subjected to the most extensive investigation. When studying techniques in connection with remote surveillance of the atmosphere, the following aspects are of importance:

- Measurement of atmospheric pollutants
- Measurement of the mechanisms responsible for the spreading of pollutants in the atmosphere: mean wind and turbulence
- Measurement of the parameters describing the dynamics of the atmosphere on which the spreading model relies: temperature distribution, precipitation.

These will now be summarized.

6.3.1 Atmospheric Constituents; Air Pollution

A laser has many advantages over a conventional thermal source of radiation for spectroscopy. It is highly monochromatic, it is coherent and highly directional, and its spectral brightness is greater by a factor of about 10^{16}. In the past the laser has suffered from a fatal disadvantage for many purposes—it could not be continuously tuned.

Various types of tunable coherent sources are now becoming available. In the visible and near infrared there are dye lasers, parametric oscillators and semiconductor diode lasers that cover the spectrum to about 30 μm. It is possible to mix the outputs in nonlinear media and thereby extend the frequency range. The spin-flip Raman laser, operating at 5 and 10 μm, is an example of a tunable laser. It is excited by a CO or CO_2 pump laser, the energy from which is directed into a cooled InSb sample. The radiation is scattered by a Raman process in which electrons reverse their spins in a magnetic field. The sample forms a cavity so that the Raman scattering takes place coherently with efficient

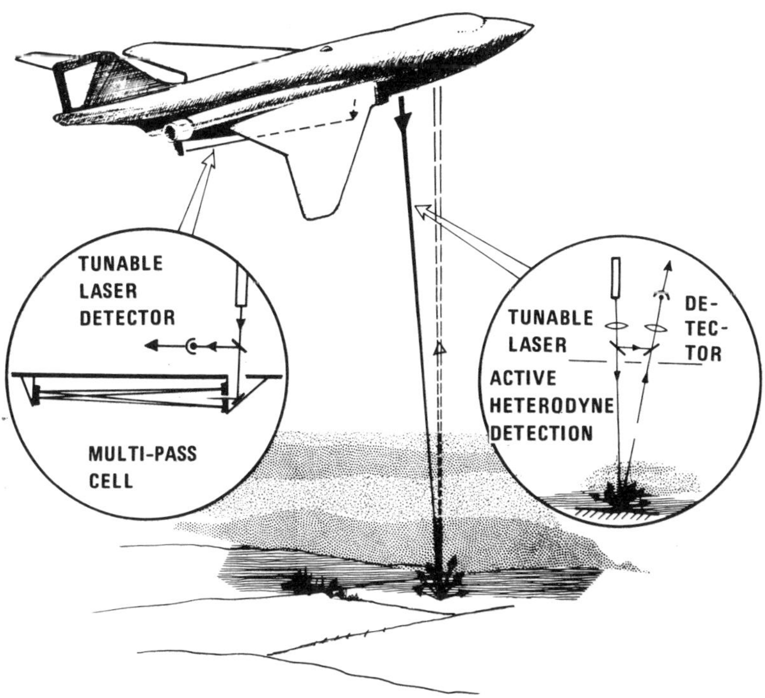

Figure 6.11 Measurement of air pollution by the use of an aircraft.

energy conversion into the scattered beam. The output frequency is controlled by varying the magnetic field in the sample.

6.3.1.1 A Brief Description of Some Important Spectroscopic Techniques (Dr. E. D. Hinkley, JPL, Dr. R. T. Menzies, JPL, Dr. F. Allario, NASA Langley)

1. *The differential absorption method* is presently one of the most powerful means of measuring small concentrations of particular pollutants averaging over an appropriate path that may be approximately a kilometer. There are several methods on the basis of

which the absorption coefficient, and hence the total gas concentration over a given path, can be measured:

- A pulsed laser can be used. The laser transmits wavelengths λ_1 and λ_2 alternatively. Wavelength λ_1 is in the absorption band, while λ_2 lies outside this band. At the receiving site (bistatic arrangement or a monostatic arrangement making use of a retroreflector) the ratio of the power received at the two frequencies $P_R(\lambda_2)/P_R(\lambda_1)$ is measured. Today it is possible to stabilize the laser system so as to be able to measure variation in $P(\lambda_1)/P(\lambda_2)$ down to 0.3%. The limiting factor with regard to detection sensitivity is then constituted by atmospheric turbulence in the absence of interfering gases. An example illustrating the potential of this technique is shown in Figure 6.12 [55-57].

2. *The derivative technique* is based on a frequency modulated laser (usually a diode laser). The laser frequency is centered on the slope of the appropriate absorption curve or line. A small frequency deviation allows the slope $dP/d\lambda$ to be measured. From this the concentration is derived. Using a high sweep rate (modulation frequency) and real time compensation of extinction fluctuations by monitoring the received power, the effect of turbulence can be minimized.

Typical sensitivities obtained with 1-km path length are as follows [58,59]:

Gas	CO	O_3	NO	SO_2	C_2H_4
Sensitivity, ppb	1.3	3	1.5	15	0.8

3. *The spectrophone* (Figure 6.13) measures the gas concentration at a point in the *in situ* manner. It is often referred to and should therefore be included in this survey. The laser is normally pulsed. The pulsed laser power gives rise to pulsed heat dissipation in the absorbing gas contained in the enclosure (shaded area on the figure). These heat pulses cause pressure variations that are measured by the microphone. Optimum sensitivity is achieved by a coherent detecting process. The advantage of this system relative to many other spectroscopic methods is that the sensitivity is proportional to the laser power. Typically 100 mW of power is used, giving parts per billion sensitivities.

The spectrophone appears to be a simple, noncritical instrument that presents an efficient solution to many problems. Whereas it was previously claimed that an acoustically resonant system is

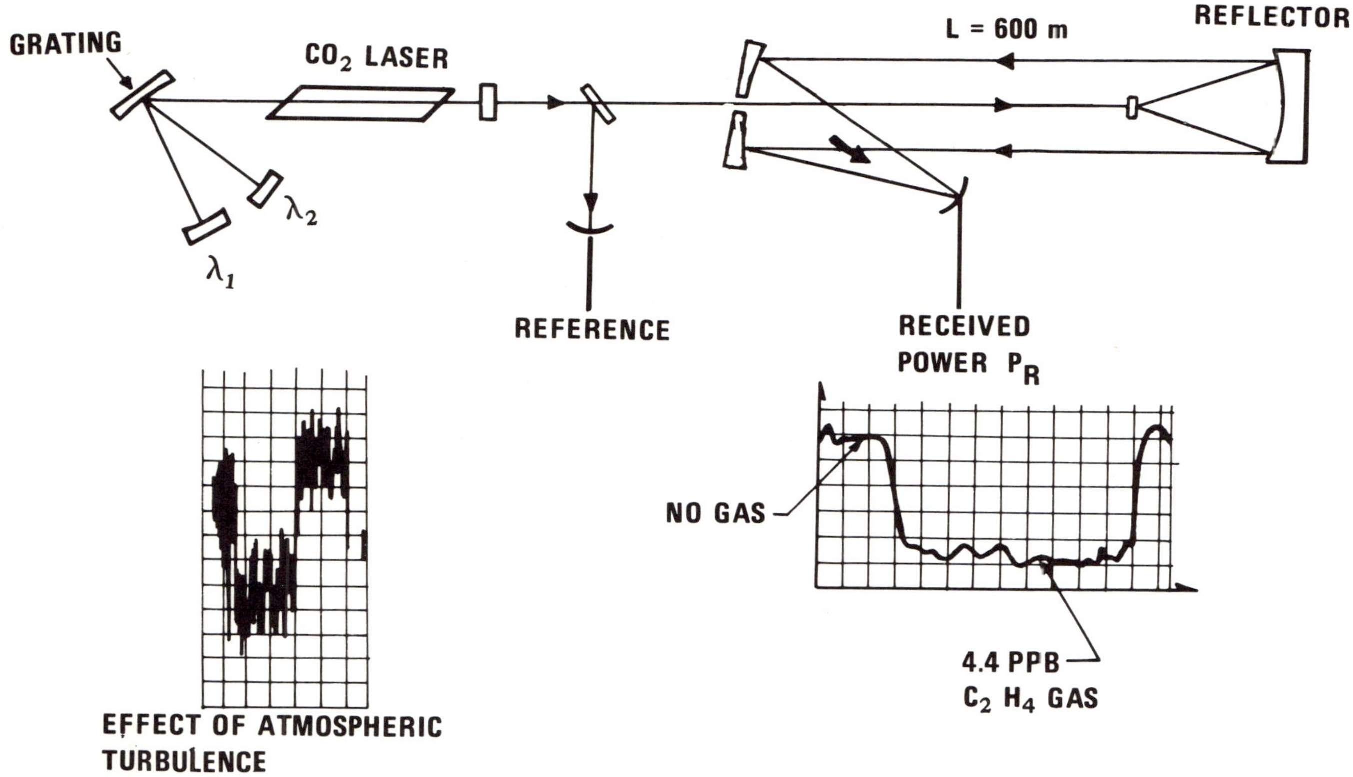

Figure 6.12 Results obtained with a dual wavelength CO_2 laser at the Norwegian Defence Research Establishment. S. O. Kanstad *et al.* [57].

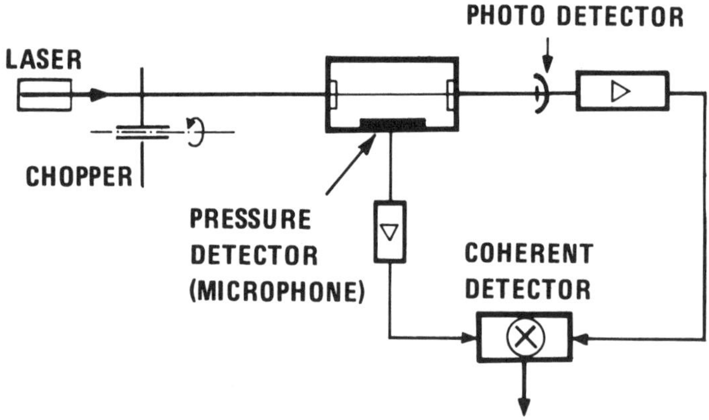

Figure 6.13 The principle of operation of a spectrophone

essential for high sensitivity, there are now many highly sensitive, nonresonant systems operating with a variety of simple microphone configurations [58-62]. The spectrophone will probably have an even greater impact on the spectroscopy of solids and liquids. Powder samples may be used and information about thermal diffusivity may be obtained by pulse rate response studies [63].

4. *Heterodyne receiving techniques* offer great improvements in system sensitivity relative to that of direct detection (order of 10^6), in particular in the infrared. Considerable progress is being made in heterodyne laser detection techniques. Limiting factors appear to be the stability of the local oscillator if this is to be a tunable one, in addition to various local oscillator noise effects. With fixed frequency local oscillators, a mixing efficiency of 0.5 is obtained; with less ideal sources (*e.g.*, a tuned diode laser) a 10 dB loss may be incurred. NASA Langley Research Center (Dr. Allario), in cooperation with the Airborne Instrument Laboratory in Melville, New York, has measured the vertical profile of atmospheric NH_3 from the ground to 15 km using a dual photomixer, dual local oscillator, and Infrared Heterodyne Spectrometer (IHS). The photomixers have a measured total bandwidth of 1.5 GHz, and a multi-IF channel, Dicke-switched output. Fixed frequency CO_2 lasers using the $C^{13}O_2^{16}$ isotope were used as local oscillators,

and the instrument was used in a solar pointing mode. The IHS was also used with waveguide $C^{12}O^{16}$ laser local oscillators to perform measurements of the vertical distribution (10-40 km) of stratospheric O_3 from an aircraft platform, the NASA CV-990. Preliminary measurements have been reported [64] and final reduction of the data is being compared to correlative O_3 chemical measurements from simultaneous balloon launches. Laboratory measurements are currently being performed at NASA Langley Research Center to study degradation factors in the signal-to-noise ratio and mixing efficiency when utilizing tunable diode lasers as local oscillators. Measurements on heterodyne noise effects induced by mode competition and cryogenic cooler instabilities are currently being studied [64-67].

6.3.1.2 Some Examples of Practical Results Obtained

1. *SO_2, N_2O, O_3; lidar techniques (J. M. Hoell, NASA Langley)* The differential absorption technique using a UV laser system and utilizing natural atmospheric particles as reflectors appears to have a great potential. NASA Langley uses a commercial flashlamp-pumped, frequency-doubled dye laser, a 0.25-m diameter receiver telescope, and a minicomputer for control, data acquisition and processing. The laser output is typically 100 μJ per pulse at 15 pulses per second for wavelengths coincident with the peaks and valleys in the SO_2 absorption spectrum from 296.0 to 300.1 nm. The difference in absorption coefficient for each pair of transmitted wavelengths is monitored using a satellite beam from the laser. The system is directed in a manner so as to record the SO_2 emitted from a steam plant in the neighborhood With the wind blowing in the appropriate direction, a suitable pollution source is obtained.

Through the use of near simultaneous pulses, the sensitivity of the system when averaging 100 pulses at each frequency in succession is about 10 ppb over a distance of 0.8 km. The system sensitivity is expected to be improved so as to give a 4-ppb sensitivity over a 1-km path. The only interferent gas is O_3, and the effect of this is small. See Figure 6.14 [68-70].

2. *Gases, CO_2 laser (Dr. R. T. Menzies, JPL).* A compact waveguide CO_2 laser system has been assembled at JPL for the

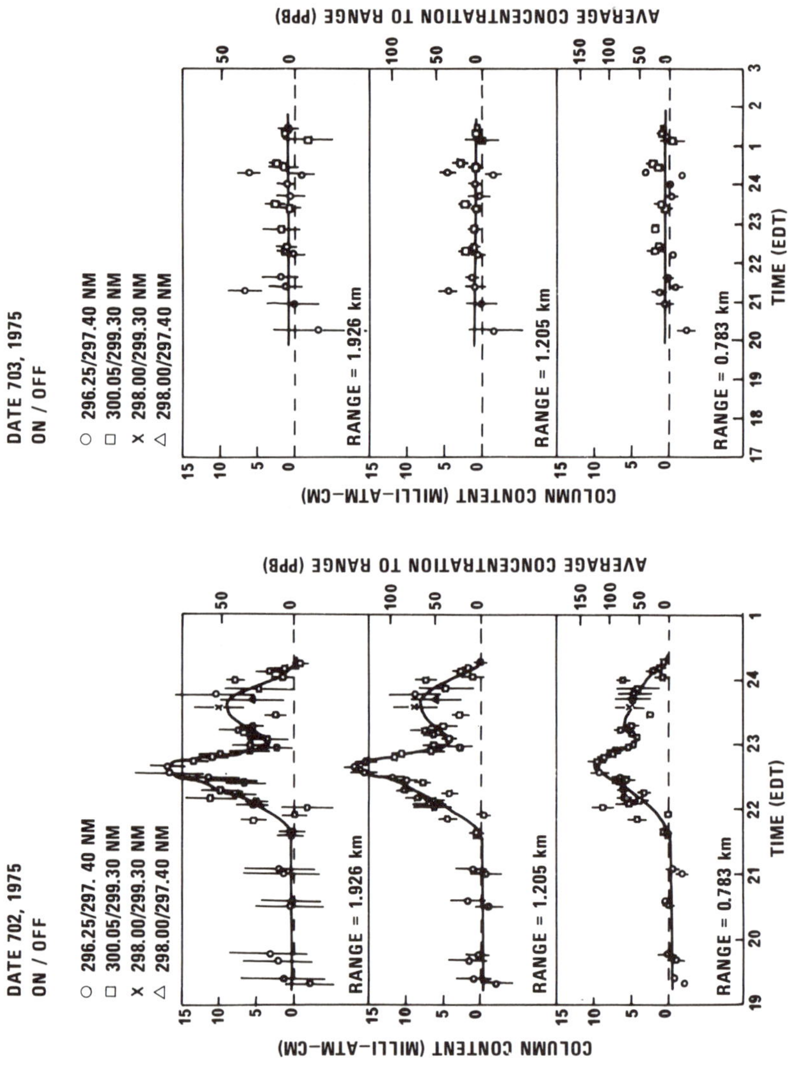

Figure 6.14 Time history of atmospheric sulfur dioxide concentration along measurement path to three ranges. After J. M. Hoell *et al.* [68].

measurement of O_3, C_2H_4 and NH_3, among others, using the ground as the reflector. The laser has been vibration-tested and clearly is remarkably robust (developed by Dr. R. L. Abrams, Hughes Research Laboratory). By means of this aircraft-mounted system the atmosphere will be measured using the ground as reflector. The system, which is based on heterodyne detection, is expected to measure variations in atmospheric absorption (measured signal) down to some 3% [71]. This figure is to be compared with the 0.3% obtained by Hinkley, MIT Lincoln Laboratory, using a ground-fixed installation with a retroreflector and diode tuning [58].

A system based on two separate CO_2 TEA lasers is being developed for an airborne application. The laser will be fired downwards from the aircraft. By using the ground as reflector, the total burden of a given pollutant (O_3) between ground and the aircraft is measured (Dr. Melfi, EPA):

- by direct detection method
- by transmitting the two frequencies in succession 12 μs apart in time
- by measuring the ratio of the power received at the two frequencies. Note: no averaging is performed before the ratios are taken.

It is very probable that this system will be highly susceptible to the geometrical structure of the reflector, the ground. Since the power at the two frequencies is compared almost instantaneously in time (12 μs apart), the effect of turbulence is not likely to be important. A cassegranian telescope receiving system with a heterodyne detector based on a CO_2 waveguide laser is available commercially (Honeywell, Inc.).

3. *Particles; temperature inversions; laser scattering (Dr. J. A. Eckert, EPA).* EPA, Las Vegas, use an aircraft-mounted pulsed ruby laser looking downwards. The Lidar receiver echoes scattered from particles in the air and from the ground. This method localizes the boundary layer (temperature inversion) by increased backscatter due to its trapping of atmospheric particles. The system is also able to track plumes originating from industrial areas. The laser is pulsed at a rate of 1 pulse every 20 seconds. The spatial resolution is about five meters. A very detailed structure (isolines of reflectivity) is obtained; these look very similar

to the isoplates of temperature obtained from acoustic backscatter systems (Sodars) (see Figure 6.15). The system is bulky and heavy (600 pounds) [72].

A new compact system (90 x 30 x 40 cm) based on a *dye laser* that can be fired once every second has now been developed. This system displays the data in real time on board the aircraft, making the device particularly useful for tracking plumes.

Finally, a neodynium YAG laser is under development. This will give spatial resolution of some 15 ft and it will have facilities for two-frequency operation. This will make it possible to obtain some information about particle size distribution.

NASA Langley is investigating techniques for the measurement of *particle size distribution* up to a height of some 20 km in the atmosphere. The experimental scattering cross section versus frequency relationship shows a fine-scale structure (interference effects) that is not fully revealed by the Mie theory (Dr. Allario).

To learn something about the mechanisms involved, microwave simulation techniques are used. Objects with shapes and dielectric constants similar to those of natural aerosols are produced (scaled from laser wavelengths to microwave wavelengths). The results of these scatter model experiments are scaled back to practical laser cases. These simulated results are then compared with the results from atmospheric laser experiments. Based on this method the atmospheric particle size distribution is synthesized.

4. *Stratospheric hydrochlorides (C. R. Farmer and Associates, JPL).* Farmer uses a high resolution Fourier transform interferometer spectroradiometer (FTS) in the 1800-5600 cm^{-1} with a spectral resolution of 0.13 cm^{-1} [73].

An aircraft is used as the platform at 60,000 ft and limb scanning is performed (Concordes in cooperation with UK and France, and U2 in the USA). Measuring the profile of HCl from the ground and up to 21 km revealed that the concentration decreased from approximately 1 ppb near the ground to a very much smaller value at the height of the tropopause. From this height the concentration increased again and reached 0.6 ppb at 21-km height. Later balloon experiments have confirmed this. At 37 km the observed mixing ratio is 1.7 ppb.

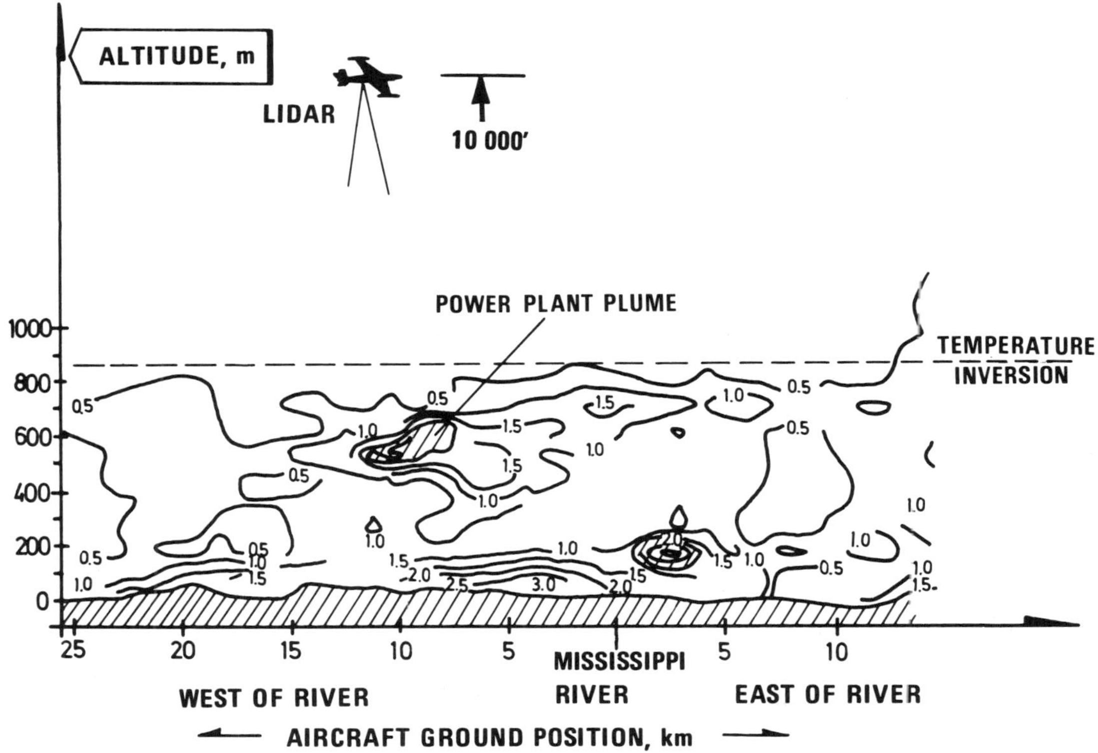

Figure 6.15 Isoscattering contour plot. After Eckert, EPA [72].

6.3.2 Atmospheric Wind

Remote measurement of the atmospheric wind field has been the subject of investigation at many research establishments for more than a decade. In addition to serving the pollution transport modellers, wind probing techniques are of importance in aviation. Consequently the effort in this field of science is considerable and the number of different techniques of significant potential is large. In this brief report only a few will be mentioned.

1. *Laser Doppler Techniques*

The NASA-MSFC has developed laser Doppler systems for remote measurement of atmospheric wind velocity and turbulence. These systems utilize the Doppler frequency shift of the laser light when it is scattered by minute particles suspended in the atmosphere. The velocity is determined directly from the measurement of the difference in frequency between the scattered and reference laser energy and the geometry of the system. The Doppler frequency is directly proportional to velocity. The feasibility of

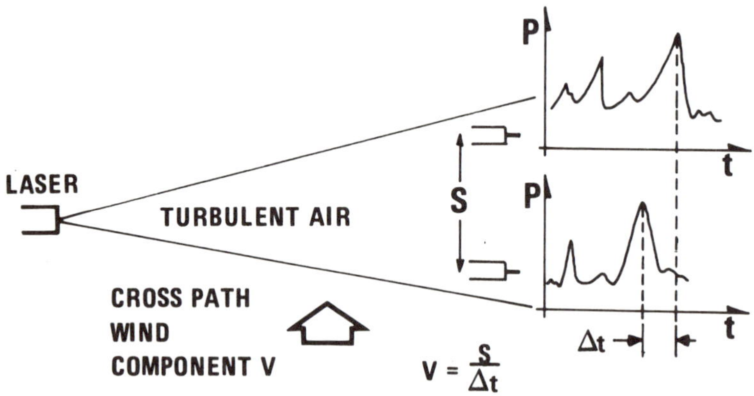

Figure 6.16. The principle of laser scintillation anemometer. The refractivity irregularities blown through the beam cause the laser signal to scintillate. The spatial correlation of these fluctuations gives information about wind speed.

measuring atmospheric wind velocity and turbulence, using a CO_2-laser Doppler system, was demonstrated at NASA in 1968. Since 1968 this technology has been directed toward (1) the development of a laser Doppler system for the detection and measurement of clear air turbulence in an onboard airline warning system, (2) the development of laser Doppler systems for detection, tracking and measurement of aircraft tailing vortices, and (3) research measurement of particular atmospheric wind fields.

Lidar (or "laser radar") provides a unique technique for observing the temporal and spatial variation of the atmospheric aerosol. Both particles and the atmospheric gases elastically backscatter pulsed laser energy at optical or near-optical wavelengths and provide information on the presence, location, and, with qualifications, the concentration of the scatterers.

Such observations are valuable in their own right, showing the structure of the atmosphere as revealed by the degree and variations of its particulate loading. Lidar measurements can also be valuable in conjunction with other observations concerned with radiative energy transfer in the atmosphere—especially of the passive type—since they provide data to complement and support information obtained on the basis of the radiometric instrumentation alone [75].

An *IR Doppler System* has been developed by Dr. R. L. Schwiesow and others of WPL, NOAA. The radial component is measured with a CW-CO_2 laser. The signal backscattered from aerosols is mixed with an infrared local oscillator. The heterodyne beat frequency is proportional to radial wind speed. Range resolution is achieved by beam focusing of the 30-cm coaxial optics. The range resolution is 10 m at 100 m and degrades as the square of the range. Angular resolution is 0.1 milliradians. Transmitted power is 3 watts CW. A spectrum analyzer yields a complete velocity-component spectrum as often as every 10 ms.

A system mounted in a light truck has been used for boundary-layer wind profile measurements to a range of 1 km. A similar system in a light aircraft has been used for the study of waterspouts and cloud dynamics. Closely-spaced multiple beams have been used to demonstrate the measurement of all three components of the velocity at a single point in space [76].

2. *Optical Scintillation Methods*
(Dr. R. S. Lawrence, WPL, NOAA)

The present methods involving line-of-sight measurements with two transversally spaced receivers and cross-correlation measurement give essentially the path average cross wind. The system may either be a one-ended system using a naturally illuminated scene (hillside, forest or buildings) as the source, or a double-ended system with a laser as the source. In the passive mode (one-ended system) the two-dimensional spatial spectrum of the scene, which in general is not the same as the white spectrum of a point laser source, has to be specified [77,78].

3. *Scintillation Method Using Spatially Filtered Apertures*
(Dr. R. W. Lee, Dr. A. T. Waterman, Stanford)

The idea of this wind-measuring technique is by the use of spatially filtered apertures to measure the transverse wind velocity as a function of the position along the path. Using two spatially filtered apertures (both transmitter and receiver) it can be shown that the observed variance at receiver output is associated with the intensity of a particular Fourier component of the refractivity irregularities located at a known distance from the transmitter. Measurements of the fluctuation frequency then lead immediately to the determination of the wind velocity through the use of Taylor's hypothesis ($2\pi f = VK$). The available space resolution is high, approximately 1% of the path length, which again may range from 10 m (optical and acoustic frequencies) to tens of kilometers (millimeter waves) [79].

4. *FM/CW Radar System*
(Dr. E. E. Gossard, Dr. R. B. Chadwick, WPL, NOAA)

Great advances have been made in this field largely due to new and very effective data reduction methods which give both the distribution in space of the scattering cross section (scattering caused by irregularities in refractive index) and the distribution in space of wind (systematic as well as the turbulent component, the variance). The technique for separating the Doppler information from the range information in short is as follows.

The signal from the homodyne mixer (frequency contains information on range + Doppler) is A/D converted. The Doppler velocity spectrum (Doppler shift as well as Doppler broadening) is obtained from each resolution cell by computing the power spectrum from a sequence of consecutive sweeps. Target range is obtained from the signal frequency by means of the usual FM/CW technique. The technique offers a unique solution to the problem of all-weather velocity measurements in the atmosphere up to a height of some 2 km.

The vertical profile of mean wind and of the wind fluctuations can be obtained along any line in the atmosphere within the region of the radar (hemisphere of radius 2 km). Days with little refractivity variations present a problem. In order to take care of such conditions an extra 20 dB sensitivity is desirable [80].

5. *Meteorological Doppler Radar for Measurement of Storm Structure by Measuring the Raindrop Distribution*
(Dr. E. E. Gossard, WPL, NOAA)

Two radars are used to obtain direction of particle motion. Chaff has been used with great success in daytime with convective activities holding the chaff up and causing good mixing and spreading of the chaff. Recent developments indicate that the system may operate also in clear air conditions [81].

6. *Acoustic Radar Techniques*
(Dr. F. F. Hall, WPL, NOAA)

The acoustic Doppler radar (Figure 6.17) is also a powerful tool. By this, wind velocity and wind variance profiles can be obtained up to a height of some 500 m. Under conditions of high wind speed over the antenna, acoustic noise and drift of the acoustic wavefront are a hindrance. Wind shear investigations are routinely carried out at Denver Airport using the technique [82].

6.3.3 Atmospheric Temperature Distribution

In connection with the study of the dynamics of the atmosphere, information about temperature distribution (vertical profile) is imperative. Consequently, considerable effort has been invested to establish remote probing techniques. This has proved to be difficult. Although, as we shall see in the following, remote probing techniques have been developed, they are at present

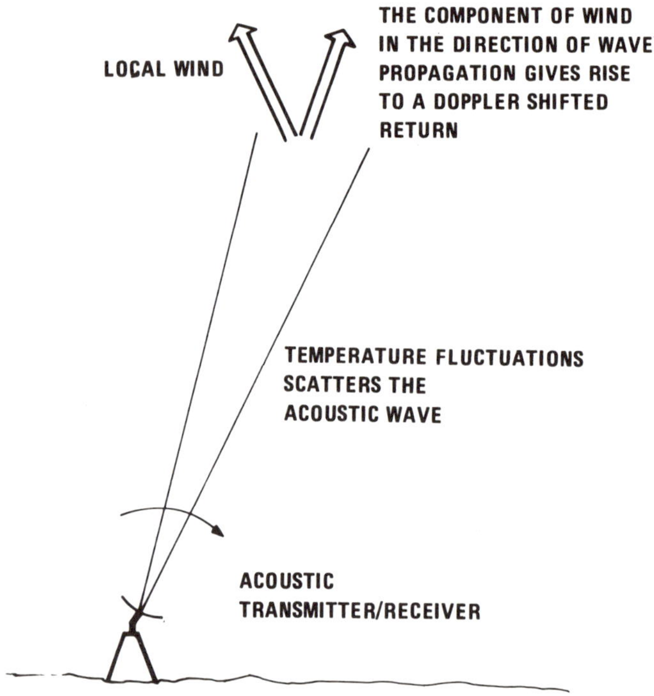

Figure 6.17 The acoustic Doppler radar for measurement of wind speed.

considered inadequate because of insufficient height resolution. However, a few methods should be mentioned.

1. *Passive Radiometry*

A passive microwave radiometer is tuned to the frequency corresponding to an absorption line (emission line) of a particular gas, such as O_2. The signal that the radiometer receives (ground-based or placed in an aircraft or a satellite) is determined by the temperature of the gas and its concentration. Knowing the concentration, which in general is the case with O_2, enables deduction of the temperature. However, our interest is not the mean temperature through the whole depth of the atmosphere, but the temperature profile. Therefore, one will have to make use of another characteristic feature of the radiating gas. The width of the frequency spectrum (line width) that a given gas emits is determined by the pressure and the molecular temperature. Accordingly, if the emission is measured at a set of closely spaced frequencies around the emission line, the temperature profile may be obtained by solving the integral equation (inversion technique) provided the pressure profile is known. In practice, however, this does not give very good height resolution due to lack of knowledge of precise pressure profile [83].

2. *Active Raman Backscatter Technique*

The Raman backscattered intensity is proportional to the molecular density of the particular gas (*e.g.*, N_2). The backscattered power in the Raman process is distributed over several rotational and vibrational emission lines. The relative strength of these lines is determined by the temperature of the N_2 molecules. Hence, measuring the distribution of line intensity, the temperature can be obtained. Unfortunately, however, also this method suffers from comparatively poor spatial resolution and poor sensitivity.

3. *Radio Acoustic Sounding System (RASS)*
(Dr. A. T. Pedersen, Stanford)

An acoustic grating whose velocity is determined by air density (temperature dependent) is "'chased" by a radio wave with wavelength twice the acoustic wavelength. The RF wave is scattered back (Bragg scattering) and the Doppler shift is measured. Thus the velocity of the acoustic wave is measured and hence the temperature. This method suffers from destruction of the acoustic phase front as a result of wind, but is in other respects effective.

REFERENCES

REFERENCES

1. Gordy, W. "Microwave spectroscopy," *Handbuch der Physik* (Berlin: Springer-Verlag, 1957), p. 28.
2. Ghosh, S. H. and V. Malaviya. "Microwave absorption in the earth's atmosphere," *J. Atmos. Terr. Phys.* 21:4 (1961).
3. McClatchey, R. A. *et al.* "AFCRL atmospheric absorption line parameters compilation," AFCRL-TR-73-0096, ERP434 (1973).
4. Ratcliffe, J. A. "Some aspects of diffraction theory and their application to the ionosphere," *Rep. Progr. Phys.* 19:188 (1956).
5. Gjessing, D. T. "Determination of permittivity variations in the troposphere by scatter-propagation methods," *Proc. IEE Monograph* 501 E, April 1962, *Proc. IEE Part C,* 109:16 (1962).
6. Batchelor, G. K. "The scattering of radio waves in the atmosphere by turbulent fluctuations in refractive index," Research Report EE262, School of Electrical Engineering, Cornell University (1955).
7. Balser, M. "Some observations on scattering by turbulent inhomogeneities," *Trans. IRE* AP-5(4):383 (1957).
8. Gjessing, D. T. "Atmospheric structure deduced from the forward scatter wave propagation experiments," *Radio Sci.* 4(12):1195-1210 (1969).
9. Wiener, N. "Generalized harmonic analysis," *Acta Mathematica* 55:117 (1930).
10. Gjessing, D. T. "On the influence of atmospheric refractive index irregularities on the resolution performance of a radar," *Atmospheric Effects on Radar Target Identification and Imaging.* H. Jeske, Ed. (D. Reidel Publishing Company, 1975).
11. Hagfors, T. "Investigation of the scattering of radio waves at metric wavelengths in the lower ionosphere," *Geophys. Norwegia* 21:2 (1959).
12. Gjessing, D. T. and J. Borresen. "The influence of an irregular refractive index structure on the spatial field-strength correlation of a scattered radio wave," *IEE Conf. Proc.* 48:September 1968.

13. Rice, S. O. "Mathematical analysis of random noise," *Bell Systm. Tech. J.* 23:282 (1944).
14. Gjessing, D. T. and F. Irgens. "On the scattering of electromagnetic waves by a moving tropospheric layer having sinusoidal boundaries," *IEE Trans. PTGAP* AP–12:1 (1964).
15. Gjessing, D. T. and F. Irgens. "Scattering of radio waves by a moving atmospheric rippled layer; A simple model experiment," *IEE Trans. PTGAP* AP–12:6 (1964).
16. Gjessing, D. T. "On the radiometeorological determination of refractive index and air velocity spectra of the troposphere," *Radio meteorologie en vue de l'application aux telecommunications*, P. Misme, Ed. (Liege: Vaillant-Carmanne Publishing Company, 1966).
17. Gjessing, D. T. "Radiophysical aspects of irregular structure in the atmosphere," *Atmospheric Turbulence and Radio Wave Propagation,* A. M. Yaglom, V. I. Tatarsky, Eds. (Moscow: Nauka Publishing House, 1967).
18. Lee, R. W. and J. C. Harp. "Weak scattering in random media," *Proc. IEEE* 57:375 (1969).
19. Kjelaas, A. Private communication.
20. Gjessing, D. T., A. G. Kjelaas and J. Nordo. "Spectral measurements and atmospheric stability," *J. Atmos. Sci.* 26(3):462-468 (1969).
21. Gjessing, D. T. and K. S. McCormick. "On the prediction of the characteristic parameters of long distance tropospheric communication links," *IEE Trans.* COM–22:9 (1974).
22. Gjessing, D. T. "On the use of forward scatter techniques in the study of turbulent stratified layers in the troposphere," *Bound Layer Meteorol.* 4:377-396 (1973).
23. Janes, H. B. and M. C. Thompson. "Comparison of observed and predicted phase-front distortion in line-of-sight microwave signals," *Trans. IEE* AP–21(2):263-311 (1973).
24. Kjelaas, A. G., P. E. Nordal and A. Bjerkestrand. "Multiwavelength scintillation effects in a long-path CO_2 laser absorption spectrometer," Proc URSI Commission F Symposium, La Baule, France, 28 April–6 May 1977.
25. Gjessing, D. T. "A generalized method for environmental surveillance by remote probing," *J. Radio Sci.* March/April (1978).
26. Gjessing, D. T. "Scattering mechanisms and channel characterization in relation to broadband radio communication systems," AGARD Conf. Proc. *Aspects of Electromagnetic Scattering in Radio Communications,* Cambridge, Massachusetts, October 1977.
27. Plant, W. J. "Studies of backscattered sea return with a CW, dual-frequency, X-band radar," *IEE Trans.* AP–25:28 (1977).
28. Alpers, W. and K. Hasselmann. "The two-frequency microwave technique for measuring ocean wave spectra from an airplane or satellite," (1977).

29. Barrick, D. E. "Remote sensing of sea state by radar," *Remote sensing of the Troposphere,* V. E. Derr, Ed. (Washington, D.C.: U.S. Government Printing Office, 1972).
30. Wilheit, T. T. and M. G. Fowler. "Microwave radiometric determination of wind speed at the surface of the ocean during BESEX," *IEE Trans.* AP–25:111-120 (1977).
31. Webster, W. J., Jr., T. T. Wilheit, D. B. Ross and P. Gloersen. "Spectral characteristics of the microwave emission from a wind-driven, foam-covered sea," *J. Geophys. Res.* 81:3095 (1976).
32. *Project plan for SeaSat–A 1978 Mission.* Project document 622-3, Jet Propulsion Laboratory, Pasadena, California.
33. Briscoe, M. G., O. M. Johannessen and S. Vincerzi. "The Malthese oceanic front: a surface description by ship and aircraft," *Deep Sea Res.* 21:247-262 (1974).
34. Blume, H-J. C. and B. M. Kendall. "Measurement of ocean temperature and salinity via microwave radiometry," *Bound Layer Met.* (1977).
35. Thomann, G. C. "Remote measurement of salinity in an Estuarine environment," *Remote Sensing Environ.* 2:249-259 (1973).
36. Smith, S. D., M. I. Colles and G. Peckham. "The measurement of surface pressure from a satellite," *Quart. J. Roy. Meteorol. Soc.* 98(416): 431-433 (1972).
37. Barrick, D. E., J. M. Headrick, R. W. Bogle and D. D. Crombie. "Sea backscatter at HF: Interpretation and utilization of the echo," *Proc. IEE* 62(6):673-680 (1974).
38. Barrick, D. E. "First order theory and analysis of MG/HG/VHF scatter from the sea," *IEE Trans.* AP-20(1):2-10 (1972).
39. Crombie, D. D. "Backscatter of HP radio waves from the sea," *Electromagnetic Probing in Geophysics*, J. Q. Wait, Ed. (Boulder, Colorado: The Golden Press, 1971), pp. 131-162.
40. Measures, R. M. and M. P. F. Bristow. "The development of a laser fluorosensor for remote environmental probing," AIAA Paper 71-1121, Joint Conference on Sensing of the Environmental Pollutants, Palo Alto, California, November 8-10, 1971.
41. Bristow, M. P. F. "Airborne monitoring of surface water pollutants by fluorescence spectroscopy," Submitted for publication in *Remote Sensing of the Environment*.
42. Kim, H. H. "New algae mapping technique by the use of an airborne laser fluorosensor," *Appl. Optics* 12(7):1454-1459 (1973).
43. Mumola, P. B., O. Jarrett and C. A. Brown. "Multiwavelength lidar for remote sensing of chlorophyll *a* in algae and phytoplankton, *The Use of Lasers for Hydrographic Studies*, H. H. Kim, Ed., Proceedings of a symposium held at NASA Wallops Flight Center, September 12, 1973.

44. Browell, E. V. "Analysis of laser fluorosensor systems for remote algae detection and quantification," NASA TN D-8447 (June 1977).
45. Pilon, R. O. and C. G. Purves. "Radar imagery of oil slicks," *Proc. IEE* AES-9:630 (1973). "Remote sensing of southern California oil pollution experiment, US Coast Guard Office of R&D, report July 1971.
46. "Development of a prototype airborne oil surveillance system." Final report May 1975, Prepared for US Coast Guard by Aerojet Co.
47. Axelsson, S. "Remote sensing of oil slicks," Report RL-0-3 R27, SAAB/ SCANIA AB (May 1975).
48. Nilsson, J. and T. Hagman. "Detektering av aljeutsläpp med radar," FOA report (1975).
49. Fontanel, A. and A. Roussel. "La détection des nappes d'hydrocarbures sur la mer," Report from Institut Francais du Pétrole/Centre National pour l'Exploitation des Océans (1976).
50. Hollinger, J. P. "The determination of oil slick thickness by means of multi-frequency passive microwave techniques," NRL Memorandum Report 2953 (June 1974).
51. Troy, B. E. and J. P. Hollinger. "The measurement of oil spill volume by a passive microwave imager," NRL Memorandum Report 3515 (May 1977).
52. Rambie, G. S., Jr. "A practical oil sensor," *Proc. 9th Symposium Remote Sensing of Environment* 1457-1466 (Ann Arbor, Michigan: University of Michigan, April 1974).
53. DeLoor, G. P. *et al.* "Radar cross sections of vegetation canpies determined by monostatic and bistatic scatterometry," *Proc. 9th Symposium Remote Sensing Environment* 1457-1466 (Ann Arbor, Michigan: University of Michigan, April 1974).
54. Schanda, E. "Microwave investigations of snow and soil moisture in Switzerland," *Proceedings of the First General Assembly of EARSeL* (European Association of Remote Sensing Laboratories), Strasbourg, March 23-24 (1977).
55. Ku, R. T., E. D. Hurkley and J. O. Sample. "Long-path monitoring of atmospheric CO with a tunable diode laser system," *Appl. Optics* 14:854-861 (1975).
56. Byer, R. L. and M. Garburry. "Pollutant detection by absorption using Mie scattering and topographic targets as retroreflectors," *Appl. Optics* 2:1496 (1973).
57. Kanstad, S. O., A. Bjukestrand and T. Lund. "Tunable dual line CO_2 laser for atmospheric spectroscopy," *J. Physics E:-Scientific Instruments* 10:998 (1977).
58. Hinkley, E. D. "Development and application of tunable diode lasers to the detection and quantitative evaluation of pollutant gases," Final technical report to the US Environmental Protection Agency (1971).

59. Sulzmann, K. G. P., J. E. G. Lowder and S. S. Penner. "Estimates of possible detection limits for combustion intermediates and products with line-center absorption and derivative spectroscopy using tunable lasers," *Combust. Flame* 20:177-191 (1973).
60. Delaney, M. E. "A Review of early spectrophone applications," *Sci. Prog.* 47:459 (1959).
61. Krenzer, L. B. "Ultralow gas concentration infrared absorption spectroscopy," *J. Appl. Phys.* 42(7):2934-2943 (1971).
62. Rosengren, L. G. "Optimal optoacoustic detector design," *Appl. Optics* 14(8):1960 (1975).
63. Rosenswaig, A. "Photoacoustic spectroscopy of solids," *Physics Today* 28 (1975).
64. Peyton, B. J. *et al.* "Infrared heterodyne spectrometer measurements of vertical profile of tropospheric ammonia and ozone," Presented at the AIAA 15th Aerospace Sciences Meeting, Atmospheric Environment Session, Los Angeles, California, January 24-26, 1977.
65. "Inner hemispheric survey of minor upper atmospheric constituents during October-November 1976," NASA TMX-73630.
66. Hinkley, E. D. and P. L. Kelley. "Detection of air pollutants with tunable diode lasers," *Science* 171:635 (1971).
67. Menzies, R. T. "Laser heterodyne detection techniques," *Laser Monitoring of the Atmosphere*, E. D. Hinkley, Ed. (Berlin: Springer-Verlag, (1976).
68. Hoell, J. M., W. R. Wade and R. T. Thompson, Jr. "Remote sensing of atmospheric SO_2 using the differential absorption lidar technique," Presented at International Conference on Environmental Sensing and Assessment (1975).
69. Browell, E. V., L. Perillo and F. S. Mills. "Double-pulse UV dial system for measurement of SO_2," Presented at 8th International Laser Radar Conference, Drexel University, Philadelphia, Pennsylvania, June 6-9, 1977.
70. Grant, W. B. *et al.* "Calibrated remote measurement of NO_2 using the differential absorption backscatter technique," *Appl. Phys. Lett.* 24:550-552 (1974).
71. Menzies, R. T. and M. T. Chanine. "Remote atmospheric sensing with an airborne laser absorption spectrometer," *Appl. Optics* 13:2840 (1974).
72. Eckert, J. A., *et al.* "Downlooking airborne lidar studies," August 1974. EPA report, EPA, Las Vegas, Nevada (1974).
73. Farmer, C. B. "Infrared measurements of stratospheric composition," *Can. J. Chem.* 52:1544-1559 (1974).
74. Schnidler, R. A. "A small, high speed interferometer for aircraft, balloon and spacecraft applications," *Appl. Optics* 9:301-306 (1970).

75. Huffaker, R. M. "Laser Doppler detection system for gas velocity measurements," *Appl. Optics* 9:1026-1039 (1970).
76. Abshire, N. L., R. L. Schwiesow and V. E. Derr. *J. Appl. Meteorol.* 13:951 (1974).
77. Lawrence, R. S., G. R. Ochs and S. F. Clifford. "The use of scintillations to measure average wind across a light beam," *Appl. Optics* 11:239 (1972).
78. Clifford, S. F., G. R. Ochs and T-i. Wang. "Optical wind sensing by observing the scintillations of a random scene," *Appl. Optics* 14:2844 (1975).
79. Lee, R. W. "Remote probing using spatially filtered apertures," *J. Opt. Sec. Am.* 64:1295 (1974).
80. Chadwick, R. B., E. E. Gossard and R. G. Strauch. "Analysis of an all-weather system for wind sounding up to 15 km altitude," Technical Report ERL 334-WPL 41, NOAA, US Department of Commerce (1975).
81. Miller, L. J. "Internal airflow of a convective storm from Dual-Doppler Roydon measurements," *Pure Appl. Geophys.* 113:765 (1975).
82. Beran, D. W. and F. F. Hall, Jr. "Remote sensing for air pollution meteorology," *Bull. Am. Meteorol. Soc.* 55:9 (1974).
83. Westwater, E. R., J. B. Snider and A. V. Carlson. "Experimental determination of temperature profiles by ground based microwave radiometry," *J. Appl. Meteorol.* 14:524 (1975).

INDEX